AF531688

MODERN CONCEPTS IN NANOTECHNOLOGY

ENCYCLOPAEDIA OF NANOSCIENCE

Vol. V

MODERN CONCEPTS IN NANOTECHNOLOGY

By

Dr. S.K. Prasad

School of Studies of Zoology & Biotechnology
Vikram University
Ujjain

DISCOVERY PUBLISHING HOUSE
NEW DELHI-110002

First Published-2007

ISBN 978-81-8356-247-8 (Set)

Published by

DISCOVERY PUBLISHING HOUSE
4831/24, Ansari Road, Prahlad Street,
Darya Ganj, New Delhi-110002 (India)
Phone: 23279245 • Fax: 91-11-23253475
E-mail: dphbooks@rediffmail.com
dphtemp@indiatimes.com

Printed at:

Sachin Printers, Delhi

PREFACE

There has been rapid progress in nanoscience over the last few years, particularly in the area of miniaturization. The present title **Encyclopaedia of Nanoscience** is the first scientifically detailed description of developments that will revolutionize most of the industrial processes and products currently in use. This ground breaking work draws on physics and chemistry to establish basic concepts and analytical tools and thus only it provides an indispensable introduction to the emerging field of Nanoscience, Scientific research is the ultimate tool in pushing forward the limit of understanding. But, as with any tool, research is only powerful if used properly, and to its full effects. The more we learn, the more we discover connections threading through biochemical and biophysical world. In writing this Encyclopaedia the author has made every effort to present these connections in a way that will help first time students of Nanoscience understand the subject and how very relevant it is to their lives.

In the preparation of this book large number of books and research papers have been consulted. So no authenticity is claimed.

The author expresses his gratitude to Mr. Wasan and staff of M/s Discovery Publishing House for their whole hearted co-operation in the publication of this book.

The author tried hard to be accurate and upto date in statement and realises the impossibility of completely avoiding errors therefore, the author will greatly appreciate having his attention called to any questionable statement.

Author

CONTENTS

1. Introduction 1—13

2. Identifying Nanotechnology 14—32

Scope of Nanotechnology, Threats and Futures and Politics, What's in a Name? Definitions ad Infinitum, Perspectives from Science, Cold or Hot, Quantum or Not, Nano in Nanoparticles, Not Nanotechnology, Conclusions.

3. Laboratory Methods 33—57

Scanning Tunneling Microscope, Holy Grail of Atomic Resolution, Novelty of the Scanning Tunneling Microscope, Hyping of the Scanning Tunneling Microscope, Science Fiction, Technological Futures, Role of Visions, Technological and Scientific Aspects, Top-down Technique, Bottom-up Technique, Monolayer Techniques, Scanning Microscopes, Nanoelectronics, Material Sciences, Health Sciences.

4. Apocalyptic Nanotechnology 58—85

Nanotechnology, Terminology and Basic Idea, Practical Development, Theoretical Development, State of the Field at Present, Utopian Dreams and Apocalyptic Nightmares, Utopian Dreams, Apocalyptic Nightmares, Toward a More Balanced Ethical View, Specific Field of Nanotechnology, Objectives of Specific Field Nanotechnology, Objectives Ethics, Development of the Field, Ethical Problems, Surmountable Ethical Problems, Politics of Technology, A Lesson from Space Policy, Players in the Government, Influencing the Government, Industrial Policy.

5. Revolutionary Nanotechnology 86—189

Defining Nanotechnology, Nanotechnology's Place in an Age of Ages, Military and Technological Development, Lessons from the Past, Nanofuture, Current Medical Practice, Evolution of Scientific Medicine, Volitional Normative Model of Disease, Treatment Methodology, Evolution of Bedside Practice, Changing View of the Human Body, Nanomedical Perspective, Nanomedicine and Molecular Nanotechnology, Nanomedicine: History of the Idea, Biotechnology and Molecular Nanotechnology, Naturophilia.

6. Replicators and Nanotechnology 190—194

Diatom Motility, Diatom Morphogenesis, Diatom Nanotechnology, The Future.

7. Nanotechnology and Cancer 195—210

Imaging to Detect Cancer Cells, Hyperthermia to Kill Cancer Cells, Gene Therapy/Hyperthermia Combination, Hyperthermia with Dendritic Cell Addition, Drug Delivery Using Nanoparticles, Tumour Necrosis Factor and Colloidal Gold, Localized Chemotherapy, Doxorubicin, Paclitaxel, Gene Delivery using Nanoparticles, MDA-7, Discussion, Treatment of Cancer with RNA Nanotechnology.

8. Nano-network 211—216

Mend the Gap, Building a Broader Nano-network, Conclusion.

9. Nanotechnology and Dialysis 217—225

Nanotechnology, Potential of Nanomedicine in Renal Diseases, Does Nanotechnology Apply to Dialysis? Conclusions.

10. Implication of Nanotechnology 226—244

Three Routes to Nanotechnology, Potentials, Nanotechnology Scenario, Organization of the Paper, Open Source in Information Commons, Open Source Software, Open Source for Nanotechnology, Opening up Intellectual Property, Strategy of Openness, Open Source and Biotechnology, Implication of Abundance, Elementary Recycling, From Flow to Stock, Atmospheric Commons, Paper Problems, Diamonds and Water, Inclusive Entitlements, Innovation and Abundance, Constituting Quasi-commons, Quasi-commons, Stakeholder Monitoring, Coevolving Communities of Choice, Constituting Governance in Quasi-commons, Crafting Commons for Millennia, Healthy, Wealthy and Wise, Collective Enterprises, Visions, Institutional Architecture, Conclusions.

11. Genetics and Nanotechnology **245—255**

Introduction: "Eppur si muove", The Lords of Creation, Short List of "Catastrophic Risks", What is Different Today? Conclusion.

12. Nanomaterials **256—268**

Nanoparticle-catalyzed Growth of Semiconductor Nanowire, Answers from the Past, From VLS to SLS, Quantum Wires to Nanophotonics, Nanomaterials in Organic Photovoltaic Devices, Polymer-fullerene Heterojunctions, Organic-nanocrytsal Solar Cells, Dye-sensitized Cells, Tandem Cells, Coatings, Optically Transparent Conductive Coatings, Optically Transparent Abrasion-resistant Coatings, Carbon Nanotube-Based Biosensors, Bioconjugates, Future Opportunities, Layer-by-Layer Nanoassembly for Drug Delivery, Antibacterial Applications, Nanomaterials in Fuel Cells, Nanomaterials in Other Energy Applications, Environmentally Friendly Energy Sources, Nanotube-Polymer Composites for Ultra Strong Materials, Exceptional Strength, Looking Ahead, Metal Matrix Nanocomposites for Structural Applications, Improved Properties, Low Cost Solutions, Clusters and Nanoparticles, Scalable Effects, Quantum Effects, Feynman's Vision.

Index **269—275**

1

INTRODUCTION

From the development of the earliest stone tools to the must sophisticated *microprocessor*, man has increasingly, and sometimes unwittingly, shaped the world around him through the use of his technologies. These technologies impact upon all aspects of our lives. We depend upon them for the food we eat, our transport and our communications. We rely on them for clean water and an increasingly sophisticated level of health care. Whole periods of human history are labelled by reference to the dominant technology of the time—the stone age, the bronze age, the iron age, the industrial age, the computer age. We are all familiar with these terms and use them without thinking about die profound effect that each of the technologies had—both upon the societies that created them and on the planet itself.

Frequently, we are not aware of the impacts the older technologies have had on the world in which we now live—for example, stone axes were used to fell the ancient forests that once covered the UK. and created the downs and pasture that we now recognize as our "green and pleasant land". This chapter will look at one new area of technology—*nanotechnology*—and attempt to answer the question "Should we be worried—either about what it is doing now or where it is taking us?" I wonder whether a Neolithic farmer stopped to ask himself the same question while he was felling the trees to create a new stretch of farm land for himself and his family.

Before discussing the issue of where nanotechnology is taking us, and whether we should be worried about it, we must try to understand what it is. So, the first question that is to be addressed is: "What is nanotechnology?" This is not as easy to answer as one might think, because the term encompasses a huge range of activities. Some people think it is not a single type of activity at all, while others think it is just a term that 'has been invented to allow researchers to extract In amounts of research funds from government funding agencies! Nanotechnology has received enormous attention in the last 15 years, and especially so recently; even the heir to the British throne has become involved.

Some commentators and financial observers—such the finance house Merryl Lynch—have even gone so far as to suggest that the impact of nanotechnology will be so great that the term will be used to describe a new era of world economic growth. Why all the fuss? What is this phenomenon that everyone is getting so excited about? People are far too concerned about these—particularly so with nanotechnology. The prefix "*nano*" comes from the Greek word "*nanos*" meaning "*a dwarf*". Hence "*nanotechnology*" might well simply mean a technology concerned with small things. However, "'nano" has also long been used as a prefix in scientific circles to mean one billionth (using billion in its American sense of a one followed by nine zeros). So we have the term "*nanogram*" for one billionth of a gram and nanometer for one billionth of a meter.

A nanometer is exceedingly small—only about 10 atoms across. On that score, we might expect "*nanotechnology*" to have something to do with technologies that are working, for example, at the nanometer level and this is the general sense in which the term nanotechnology is used today. It is important to distinguish here between nanoscience, which is the study of phenomena at the very small scale, and nanotechnology, which implies an aim to achieve an end that is in some way "*useful*". The Royal Society/Royal Academy of Engineering Working Group on the subject adopted the following definitions:

Nanoscience is the study of phenomena and manipulation of materials at atomic, molecular and macromolecular scales, where properties differ significantly from those at larger scale.

Nanotechnologies are the design, characterization, production and application of structures, devices and systems by controlling shape and size at nanometer scale.

The scale of sizes normally discussed for the applicability of nanotechnology is usually less than 100 nm. Nanoscience, arguably, has been around since the early part of the 20th century, while the idea that there might be some technological advantages to be gained by working at the very small scale came much later. It was first put forward by the famous physicist Richard Feynman, when he gave a lecture in 1959 to the American Physical Society entitled "There's plenty of room at the bottom—an invitation to enter a new field of physics". In this lecture—which actually had almost no physics in it but was mainly concerned with the technology of making things—he explored the benefits that might accrue to us if we started manufacturing things on the very small scale. The ideas he put forward were remarkably prescient. For example, he foresaw the techniques that could be used to make large scale integrated circuits and the revolutionary effects that the use of these circuits would have upon computing. He talked about making machines for sequencing genes by reading DNA molecules. He foresaw the use of electron microscopes for writing massive amounts of information in very small areas. He also talked about using mechanical machines to make smaller machines with increasing precision.

Interestingly, for a man who went on to win the Nobel Prize in Physics in 1965 for his work in the field of quantum electrodynamics, he hardly mentioned the quantum mechanical effects associated with the making of very small things, although he did talk about using the interactions of quantized spins, a kind of '*spin logic*', which is only now being studied. Many of his predictions in that lecture have come true, and all are aspects of what we would now call "*nanotechnology*", although he did not use the term itself. Actually, the first use of the term "*nanotechnology*" was by Norio Taniguchi who, in 1974, gave a talk describing how the dimensional accuracy with which we make things has improved over time. He studied the developments in machining techniques over the period from 1940 until the early 1970s and predicted (correctly as it turned out) that by the late 1980s techniques would have evolved to a degree that dimensional accuracies of better than 100 nm would be achievable. He called this "*nanotechnology*".

Incidentally, Cranfield Precision Engineering was one of the leading companies in helping to develop machines that could actually make things to this kind of precision. You will realize from the above that all the early running in the field of nanotechnology was made by physicists and engineers who mainly thought m terms of making things more and more precisely. This means using one machine to make another, usually smaller machine to greater precision, and using machines (frequently very large and expensive machines) to make things which have incredibly precise features defined upon them. We now call this "*top-down*" nanotechnology. It has led directly to the hugely successful semiconductor and information and communications technology ((ICT) industries, with a world market size in excess of $4 × 10², currently growing at 4.8% per annum. We all use advanced microprocessors in our portable computers. These have metal lines written on them that are only 90 nanometers wide and have upwards of 100 million transistors on a single piece of silicon a few millimeters across. They are objects of mind-boggling complexity and yet they are manufactured at incredibly low cost—a few tens of dollars each.

The technologies established by the semiconductor industry are also now being applied in the manufacture of tiny micromechanical machines for sensing and actutation. These "*microelectromechanical systems*" (also known as MEMS) are finding their way into a host of applications, particularly in the automotive and medical fields, where cost and size-based functionality are key factors. We can start to see here the enormous effects that working at the very small scale is having on our world.

There were two other themes that Feynman put forward in his lecture. He envisaged the possibility of making machines that could pick up and put down single atoms. Putting particular atoms together in particular combinations would be a new way for making chemical compounds. Feynman did put forward several reasons why atomic manipulation might not work. These included the van der Waals and chemical forces that would make an atom stick to the finger picking it up, so that once picked up it would be virtually impossible to put down, let alone put it in a particular place relative to another atom. This has been called the "*sticky fingers problem*".

In 1981 Binnig and Rohrer, based at IBM in Zurich, invented the *scanning probe microscope*. This uses a very sharp metal point scanned over a surface to create images of the atoms in the surface. Incidentally they won the Nobel prize for this work in 1986. In 1989 Don Eigler used the scanning probe microscope to nudge atoms of xenon on a copper surface held at a temperature close to absolute zero to spell out the letters "IBM". Another one of Feynman's predictions had come true, admittedly under very special conditions. Eigler and his group have since done some remarkable work, mainly using the technique to explore basic physical and quantum mechanical phenomena. Jim Gimzewski at IBM used similar techniques to push single molecules around on surfaces. This kind of work with single atoms and molecules is called "*extreme nanotechnology*".

Feynman had another vision in 1959, which was of a factory in which billions of very small machine tools were drilling and stamping myriads of tiny mechanical parts, which would then be assembled into larger products. In the late 1980s another worker in California, Erie Drexler, combined these small manufacturing ideas of Feynman with another thought experiment, which had been put forward by John von Neumann in the late 1940s. This was the idea of a mechanical machine—called a "*clanking replicator*"—that could be programmed to make replicas of itself. All it would need was a supply of raw materials and a source of energy. Those replicas would make more replicas, and the result would be exponential growth in the number of the machines—until either the source of raw materials or the energy was exhausted.

Drexler combined this "clanking replicator" idea with Feynman's to come up with the concept of the "*universal assembler*", first put forward in his book "*Engines of Creation*". The clanking replicator concept is reduced to very small size through the use of mechanical components that are made on the molecular scale. The first assembler would be programmed to make copies of itself by atomic and molecular manipulation. The exponential growth would lead to billions of assemblers that would be programmed to work in concert with each other to build virtually anything required. You could sprinkle a few assemblers onto a heap of garbage and out would come a washing machine! Clearly such a technology would have enormous economic implications, both in terms of material and energy use, effects on employment, etc. Drexler has set up the Foresight Institute in California and has raised large amounts of money for nanotechnology research based on that idea. He also was the first to raise alarm bells about the possibility of the assemblers reproducing out of control and reducing everything to a "*grey goo*" of tiny machines.

There is now a new variant on this vision, which postulates the fusion of nanotechnology with biotechnology to create an assembler that is at least partly biologically based. In this case, the problem becomes one of "*green goo*" rather than "*grey goo*", but the outcome is essentially the same. This idea was subsequently hit upon by the author and screenplay writer Michael Crichton who used it in his novel "Prey", in which a set of assemblers, originally created for medical inspection and subsequently military purposes, goes out of control and starts hunting down and destroying their creators. There have been some remarkable developments in materials science and chemistry over the last 15 years or so, particularly where small size plays a big role in determining basic properties. In the field of materials science, size does indeed matter! If we take a piece of a semiconductor less than about 100 nanometers across, then the electrons in them behave differently from in the bulk. For example, the colours of light absorption and emission change.

Very small particles (*nanoparticles*) of materials like cadmium telluride are being used in applications such as the labelling of biological molecules and in new types of displays. These can be made amazingly precisely in size— say 50 nanometers, plus or minus a couple of nanometers—using reasonably standard wet chemical processes. Very small particles (less than a few hundred nanometers in size) do not scatter visible light. Good absorbers of ultraviolet light such as titanium dioxide are now being made in nanoparticulate form for sunscreens. The fact that the particles are so small means that they are invisible on the skin, while still being highly effective as UV blockers.

Very small particles also possess high surface areas per unit of mass. Oxonica, a start-up company from Oxford University, has found that nanoparticles of cerium oxide, when introduced into diesel fuel, act as oxidation catalysts during combustion. This provides improvements in fuel efficiency of up to 10% and reduces the emissions of carbon soot from the engine exhaust. If we look at other areas of materials science, we see that new forms of carbon have been discovered. Harry Kroto from the University of Sussex, together with Richard Smalley and Robert Curl, discovered the carbon$_{60}$ molecule in 1985 (and won the Nobel Prize for chemistry in 1996). It is a sphere 0.7 nanometers across that looks like a soccer ball, or the geodesic dome structure pioneered by the 1930s architect Buckminster Fuller, so they called it *buckminsterfullerene*.

It is amusing to note that if you could expand the C_{60} molecule so that it was the same size as a soccer ball, the soccer ball itself, if blown up by the same factor, would be about half the size of the planet Jupiter! The so-called *fullerenes* form a whole family of related structures that possess remarkable physical and chemical properties. If fully fluorinated, the molecules, which can then be thought of as tiny Teflon balls, form one of the best lubricants known. In 1991 Iijima discovered carbon nanotubes. These are like sheets of graphite roiled into long tubes, each one being terminated by a fullerene group. They also have remarkable properties. They can be either metallic or semiconducting, depending on the precise way in which the carbon atoms are assembled in the tube.

The metallic forms have electrical conductivities 1000 times better than that of copper and are now being mixed with polymers to make conducting composite materials for applications such as electromagnetic shielding in mobile telephones and static electricity reduction in cars. They possess mechanical properties that are many times superior to those of steel, bringing the promise of replacing carbon fibres in a whole new generation of high 'strength composite materials. They have been demonstrated in applications as diverse as supercapacitors for energy storage, field emission devices for flat panel displays and nanometer-sized transistors.

Clearly, these nanomaterials hold huge promise for the future. So, what is nanotechnology? Is it a real subject, or is it just, as the cynics would .have us believe, a mechanism for extracting funding from gullible government agencies and investors. It employs all the conventional scientific and engineering subjects in order to achieve new applications. It does this through the exploitation of phenomena in which small size is the key to obtaining an exploitable property. Secondly, it is an area of endeavour where there are real and remarkable properties that we can seek to exploit. Thirdly, and increasingly, we are starting to see convergence between different areas of nanotechnology. The size range of interest between a few nanometers and 100 nanometers is one where many interesting things happen. All sorts of physical properties change and many biological systems function at this length scale. Hence, we are starting to see the use of processes such as electron beam lithography, originally developed for writing very fine scale features on silicon for electronics, being applied for the modification of surfaces on which biological species can be grown in a controlled way.

The self-assembling properties of biological systems, such as DNA molecules, can be used to control the organization of objects such as carbon nanotubes, which may ultimately lead to the ability to grow parts of an integrated circuit, rather than having to rely upon expensive top-down techniques. This type of self assembly is called "*bottom-up*" nanotechnology. It is hoped that the engineering complexity of integrated circuits means that the *top-down methods* will be with us for a long time to come but self-assembly techniques may have an increasing part to play. We are thus seeing an area that is providing real potential in its applications. Some people have predicted a world market for

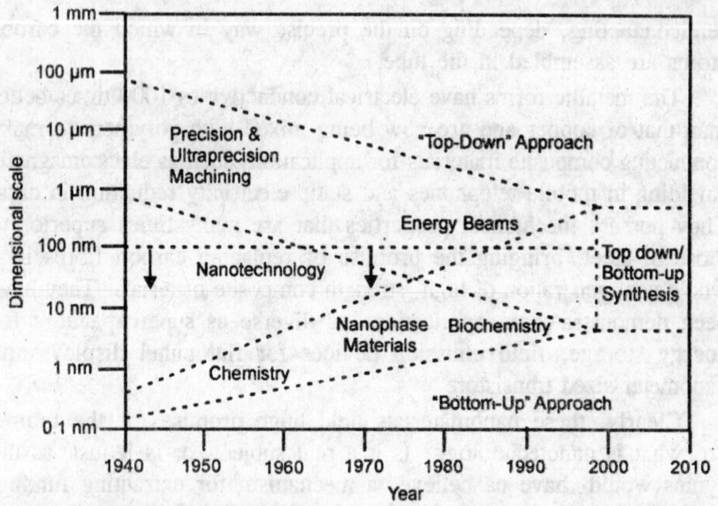

Fig 1.1. Illustrating the convergence of top-down and bottom-up nanotechnologies.

nanotechnology-related products of billions or trillions of dollars by the end of the decade. There is no doubt that the technology associated with our ability to manipulate matter on the very small scale is already having major impacts on our lives and this impact will only increase. Hence, we should now ask the question that our *Neolithic farmer* (presumably) didn't: "Should we be worried?" Could the introduction of nanotechnology have unforeseen consequences? First let's consider the *Drexlerian dystopia* in which a rogue molecular assembler, ostensibly created for the betterment of mankind via high efficiency, low cost manufacture, goes out of control and reduces everything to a *grey* (or *green*) *goo*.

There are many highly rated, first class scientific minds (including Richard Smalley, the Nobel Prize winner referred to earlier) who have asserted that the "assembler" is not possible for all .sorts of reasons. These include the "*sticky fingers*" problem, problems with the storage and transmission of the huge amount of information needed and the vast complexity of the problem, which would be far, far greater than the complexity of a modem microprocessor. They imply that the assembler concept needs to stay where it belongs, firmly in the realms of Michael Crichton's science fiction novel. There are already very real self-assemblers all around us that owe nothing to nanotechnology. They are called *viruses* and *bacteria*. They pose a very real threat, that is moreover increasing, due to our farming practices, profligate use of antibiotics and cheap international travel. The creation of this hazard obviously cannot be laid at the door of nanotechnology, although some of the nanoscale analysis techniques evolved for nanotechnology may well be able to help with combating it.

What about the less spectacular aspects of nanotechnology? Are there aspects of these materials that we need to be careful about? There is no doubt that the properties that give nanomaterials their technological exploitability might also give us cause for concern. We have already seen in the last 100 years how a single material with hugely beneficial properties can bring equally huge problems. Asbestos was very widely used in the period between the late 19th and mid-20th centuries. It has extraordinarily useful insulating properties. The high efficiency steam engines used in ocean-going liners and steam locomotives would have been impossible without it. However, we all now know how lethal a few asbestos fibres can be if inhaled, causing occupational cancers, and many people have died prematurely because of it. Is there any risk that any of the materials that are emerging from our nanotechnology labs might be building-up a similar problem for the future? There are certainly enough physical similarities between the dimensional characteristics of asbestos fibres and carbon nanotubes to cause some concern. At the moment we just don't know whether carbon nanotubes are hazardous; there has not been sufficient research. There has been a small amount of work in which rats inhaled carbon nanotubes, but the results were inconclusive.

It is important to note that in all of the applications such as in composites or displays, the tubes would be very firmly fixed in a stable structure and would therefore be unlikely to pose a threat to the general public, although we should still ask questions about how the product would ultimately be disposed of. Would there be a chance of carbon nanotubes being released into the environment? We should also question how the products would be manufactured and what the exposure might be for people involved in the manufacturing (or, indeed, for researchers using the materials). A full life cycle analysis is essential for these new materials.

What about nanoparticles? Should we be concerned about inhaling these, or indeed rubbing them onto our skins? There is good evidence that we have been exposed to certain types of *nanoparticles* in the atmosphere for millennia. Wood-based camp fires are excellent sources of *nanoparticulate soot*, for example. There is little cause for alarm in the sense that nanoparticles, *per se*, do not constitute a new hazard at low levels of exposure. However, there is good evidence that heavy exposure to carbon black is a serious industrial hazard, and that heavy exposure to nanoparticulate soot from sources such as diesel exhausts may be a cause of cancer. There is also evidence that, should nanoparticles arrive in the lungs, they do not remain there, but readily cross the barrier into the blood stream, whence they can migrate to various parts of the body, including the brain. There is, unfortunately, very little evidence to tell us what damage nanoparticles might cause there.

On the specific issue of sunscreens, it is known that the main oxides of interest are inert when present at the larger scale, but again there is a dearth of evidence about how the particles might behave when at the nanoscale. There is as yet, no evidence lo suggest that they can penetrate into healthy skin, but the fact that nanoparticles are known to have .different properties from those exhibited in the bulk must give us some cause for concern should, they be ingested. There is a need for much more research in this area. We should consider treating size as a property when dealing with chemicals in, for example, assessments of chemical or disposal hazards, just as we would deal with any oilier chemical property. There is also a need for companies to put the results of research on nanoparticle-based products into the public domain. The uncontrolled environmental release of nanoparticles is another cause for concern. Nanoparticles that are not bound to other materials may become widely dispersed in the environment and turn up in unexpected places.

However, we must also keep a sense of proportion. We have little evidence at the moment that oxide nanoparticles form a general hazard and there may be many benefits in using them. If a nanoparticulate oxide fuel additive can reduce nanoparticulate carbon emissions, then that may have significant benefits for the health of the general population. An improvement in fuel efficiency would make a positive contribution through reductions in carbon dioxide emissions and thereby a reduction in global warming. Similarly, if the availability of nanoparticulate sunscreens were to increase their general acceptability and thereby their use, then this might contribute to a reduction in skin cancer. What is certain is that we need to understand more about the health and environmental issues associated with nanoparticulate materials.

Are there other areas for concern about nanotechnology? *Nanotechnology*, like virtually all technologies, is being considered for the contributions it can make to defence. These include obvious developments such as improved electronic systems for communications, improved sensors and, especially, improved materials. Examples include composites using *carbon nanotubes* and a novel, body armour that uses *oxide nanoparticles* dispersed in a fluid and held between two flexible *Kevlar sheets*. The composite fabric is flexible, but any attempt to penetrate the sheets by, for example, a projectile or other weapon causes the fluid to rapidly set into a rigid mass that protects the wearer. The applications to flexible, wearable, protection that would protect a soldier's limbs as well as his torso are obvious. *Metallic nanopowders* can be expected to deliver more powerful conventional explosives. These developments would certainly give the armies using them a military edge, but hardly constitute the kind of doomsday scenario that we sometimes read about in the popular press or see on our video screens in movies like Terminator II. Those images owe more to science fiction based on the *Drexler vision* than a level-headed scientific analysis of the advances that nanotechnology is really likely to deliver.

Paradoxically, the broad contribution that nanotechnology can potentially make to economic development has sometimes been cited as a cause for concern, especially by some pressure groups. They point out that the promise of nanotechnology would widen the disparity in development between the rich and poor nations of the world. The rich will get richer, while the poor will be unable to benefit from the *nanotechnology revolution*. Another criticism that has been raised against the nanotechnologists is that they are promoting their research as possibly providing cures for disabilities such as deafness. Advanced cochlear implants based, on MEMS technologies can already provide profoundly deaf people with the ability to hear.' Such implants are much more effective if they are given to children rather than to adults because the child can more quickly adapt to the inputs from the implant and learn how to use it. However, the challenge issued by some deaf people is to say, "I am perfectly happy with the way I am. What gives you, the scientist, engineer or doctor, the right to say that I—or more particularly my child—-ought to be "cured" of something that I do not consider as a disability?" Is it society that needs to change its attitude to those who are different?

These questions of how we as a society choose to deploy or employ the capabilities that nanotechnology may provide are very good ones. But it strikes me that they are ethical issues that pertain more to the general capabilities that our modem technologies can deliver, rather than being specific issues concerning nanotechnology *per se*. They are, perhaps, more in the realm of the philosopher than the engineer. Nevertheless, the success of the nanotechnologists in promoting their area makes them natural targets for these ethical questions. We nanotechnologists need to make links with the people who can help us to answer them, and this is starting to happen.

One clear area of concern for the nanotechnologist is the perception of the subject by the general public. We certainly want to avoid the problems that have beset the *genetically modified* (GM) area, largely through the attempt by large companies to impose *GM crops* and *foodstuffs* on the general public, without first entering into a dialogue with civil society representatives or attempting to engage people in debate. The approach backfired badly; especially in Europe. A recent survey conducted early in 2004 under the auspices of the Royal Society and the Royal Academy of Engineering showed that few people had any preconceived ideas about nanotechnology. Those who had heard of the topic were more- or-less equally divided between those who generally thought of it in terms of a beneficial high technology (better computers and mobile phones) and those who vaguely perceived it as slightly threatening or menacing. We are certainly not (yet) in the *GM position*. This is not to say we should be complacent. Nor is it correct to say that "if only the general public understood more about our subject, they would warmly embrace it." This so-called "*deficit model*" of the public understanding of science is rather controversial. Nevertheless, scientists and engineers certainly need to work harder to get the public, and especially our children, more engaged with the subject. This is a real area for concern, but again it is a general one and not simply to do with nanotechnology alone.

In this chapter efforts have been made to show how nanotechnology—the exploitation of matter when it is deliberately structured at the very small scale—has the potential to provide huge benefits, just as any useful technology should. It is a real, very broadly based and multidisciplinary area of human endeavour and not just a token epithet that can be applied to the latest research proposal or business venture in an attempt to get it funded, although admittedly it is frequently used that way. Certainly, there are issues that should concern us. These can be specific. We need, for example, to understand more about the health and environmental impacts of our uses of nanoparticles. There are also more general concerns. Nanotechnology could provide us with a broad range of capabilities but they need to be applied in a thoughtful and responsible manner. However, these concerns are similar to those that might have been applied to any significant technology in the past. The only difference is that we are now in a position to learn from history and can try to take action before mistakes are made. It is generalized suggestion that the action should be both proportionate and based on a realistic analysis of the likely risks and benefits of nanotechnology.

2

IDENTIFYING NANOTECHNOLOGY

When Norio Taniguchi first referred to his work as "nano-technology" (Taniguchi, 1974), it's doubtful he imagined that scholars might one day be pondering the word and considering its implications on humanity, even on the fate of humanity. But this year research on the societal implications of nanotechnology accounts for nearly 10% of direct federal funding on nanotechnology in the United States: 80% of that on environmental and toxicological effects and the remaining on broader sociological studies. That's not pocket change out of a budget of over $1 billion USD. But how can you study the societal impact of nanotechnology when the term itself has not been clearly defined? In many ways, the label "nanotech" has been thrust on science and engineering externally, formed by society before solidifying as a field in its own right. This timing mismatch has interesting consequences for the shape of progress in nano-related fields and enterprises. A nanometer is one billionth of a meter or roughly equal to 10 hydrogen atoms laid side-by-side.

Many systems, from biological to chemical to quantum, exhibit fundamentally new behaviour at this length regime, often for completely coincidental reasons. For example, the blood-brain barrier often fails to filter particles of nanometer size, while being very selective about larger objects. The nanometer regime also happens to be where the quantum properties of electrons start to dominate in miniaturized circuits. The two have no connection in cause, but they are both considered "new nanoscale phenomena." This *reality of nanotechnology* contrasts with the *vision of nanotechnology* promoted in recent decades by futurists, science fiction writers, and doomsayers, as well as actual scientists. But *the reality* and *the vision* of nanotechnology have become entwined culturally despite their almost extreme differences.

Growing excitement in the field often traces back to two people: physicist Richard P. Feynman and researcher K. Eric Drexler. Physicists often justify their work by bowing to a past physics god, of which Feynman is one of the most famous. In his 1959 talk *There's Plenty of Room at the Bottom: An Invitation to Enter a New Field of Physics*, Feynman laid out the vision of what would one day be called nanotechnology. Simultaneously simplistic and prescient, Feynman explored the vast technical applications that mastering matter on a small scale would create. He gave crude but surprisingly insightful suggestions about how to access and take advantage of all the room "at the bottom." Although one must be careful of fulfilling Matthew's Law (automatically attributing good ideas in a field to the most famous person in that field), Feynman's talk anticipated manufacturing techniques such as the focused ion-beam etching Taniguchi was later to work on (and which is only now becoming mainstream in university labs), miniaturization in information storage, computation and machines, electron spin electronics (or spintronics), and even hinted at quantum computing.

Drexler is no physics god. But he deserves substantial credit for popularizing and motivating nanotechnology with his 1987 book *Engines of Creation: The Coming Era of Nanotechnology* and later contributions. He falls squarely under the vision category of nanotechnology. The vision of nanotechnology could be described as atomic level systems engineering. Atoms, like so many Lego blocks, can be arranged one-by-one to build microscopic analogues of macroscopic machines or indeed anything one could imagine, from scratch. This is the life analog, the construction block view of both Feynman and Drexler, whom the latter calls machine-phase nanotechnology. From drugs designed from the bottom up to artificial nano-machines that could flow through and oxygenate blood on demand, the vision is dramatic. Drexler imagines assemblers, i.e. nanomachines that build other nanomachines, able to tackle problems from the size of the body to the globe. Take wood, an undisputedly useful building material "*grown*" by nanoscale/microscale machines. What else could be "grown" if we design the life that grows it?

The implications, like those of biotechnology, can be scary. Fortunately, the vision is hard, and the reason this is so is because reality starts to intrude. The novel physics, surface science, and chemistry that begin to emerge in the nanoscale regime makes the construction block paradigms fall apart. The reality is actually much more interesting. Feynman himself pointed out that quantum effects and the complex behaviour of many interacting particles - be they electrons, atoms, or molecules - would make this building block strategy difficult. In essence, that difficulty is nanotechnology today. But even he couldn't have imagined how big and how interesting those complications would be...and more importantly, how useful. Let us pause for a brief moment to discuss bias and definitions. Many scientists could claim the authority to explain nanotechnology - biologists, chemists, materials scientists, and physicists. And all would necessarily do an incomplete job, a factor this chapter too will suffer from. Our bias is that of condensed matter and materials physics, the fields from which the transistor, solid-state laser, and silicon integrated circuit were born.

Complex terms and definitions, unfortunately, will be commonplace in such a broad discussion of new technologies and their categorization. Get used to being ignorant in this field. Brief definitions will be provided where possible but understand that scientists can use one word - say "spintronics" - to represent thousands of technical papers that may cover multiple subfields and have subtle meanings. Part of our challenge is that the forward line of the advancement of human knowledge has become so long. With so many individual innovators, complete categorization becomes impossible. You are encouraged to sleuth on your own when you encounter a cool sounding word you don't understand. Call it the joy of finding things out.

Scope of Nanotechnology

There are many reports of varying quality that describe examples and applications of nanotechnology, from the perspectives of length-scale, medicine, environment, risk, society, and public policy. This chapter will describe the social construction of this term without becoming encyclopedic. If the social construct of nanotechnology encompasses both the vision and the reality, one must consider the scope of both. Oddly, the vision of nanotechnology includes both the fantastic almost science-fiction visions of nano-assemblers and artificial life but also very non-nano visions including small machines and devices. Compared to a nanometer biological "*machines*" can be relatively huge; a red blood corpuscle is approximately 7000 nm while bacteria is 1000 nm and a virus is typically 60-100 nm. And key body dimensions are downright enormous. Swallowing a millimeter-sized pill that can take pictures of your insides, though it might be called nanotechnology, may have no nano-fabricated components whatsoever. This imprecision is present in both perspectives.

In many ways the reality of what's going on in nanotechnology today – as presently named – is broader than even the vision. At a recent conference on quantum nanoscience in Australia, a physics professor stood up and said something representative of many scientists' viewpoint: "Can't we just call it novel-technology?" Maybe we should. In the policy-technology world nanotechnology has become a catchall for all advanced materials technologies, from biointegration to molecular transistors to quantum mesoscopics. The USA National Nanotechnology Initiative is an example of this, as its proponents seek to make nanotechnology a unifying theme across many of the sciences that deal with fabrication and manipulation of matter at small-scales.

Despite these facts-on-the-ground, some have claimed that atomic and molecular engineering alone is nanotechnology. It's simply too late for this narrow and somewhat boring definition. Additionally, there are convincing reasons to lump larger systems under the nanotechnology roof. All the knowledge gained through the study of mesoscopic physics and complex systems might one day help the atomic-level construction technology. The science and technology community has to deal with putting these systems together and with the complex emergent behaviour that implies. What quantum mesoscopics, for example, gives us is a physical window on nature's immense complication.

It would also be easy to say that the vision or molecular assembly view, as theorized by Drexler's so-called machine-phase nanotechnology, is the "true" nanotechnology and stop right here. But most of the money being spent on nanotechnology and most of the people who believe they are working on nanotechnology are in the reality camp. The glow that the term nanotechnology exudes not only attracts new students to science but also attracts scientists to adopt the term and become "nanotechnologists." And it helps with funding. Will nanotechnology just become a rallying cry or does it have any real meaning anymore?

From a societal standpoint, the consequences of this emerging collection of new artifacts, both directly as threats to the environment and our health, and indirectly via the transformative properties of life-changing technologies, have unquestionable importance. But there are limits to the extent of nanotechnology's embrace. In some situations, e.g. federal funding or patent decisions, more precise categorization can become necessary. For example the patent office makes a distinction that a new invention just can't be smaller to be patentable - there has to be something new there (like a fundamentally new manufacturing process). Many of the suggestions Feynman made just took advantage of making things smaller and were not necessarily based on new phenomena at that length scale. In the research world some inventions or lines of discovery, though they might coincidentally take place at the nanoscale, are better classified in a different scientific context. These are often one of disparate yet specific subfields.

The reality of nanotechnology has matured at its own rate as technological manufacturing and measurement techniques have steadily improved over the last 50 years, putting fabrication and control of the nano realm within reach starting some 15 years ago. It's easy to realize that "seeing" was the key to progress. Indeed, improvements in microscopes such as the electron microscope to the creation of new classes of devices such as atomic force microscopes (AFMs) and many other techniques contributed to advances in small science. Much of this is just the natural progress of science and technology, driven by human curiosity, economic interests like the semiconductor or drug industries, and luck.

The construction block view of nanotechnology is divorced from the daily interests of most scientists and so far away as to be utterly ignorable. In some sense it is a very simplistic idea. "Hey! Let's design things from the ground up!" Let's just make the nanoscopic equivalent of a macroscopic device. Replace the surgeon and scalpel with the mini-robot. It might be more difficult to do, for various reasons, but you are not necessarily using any new science to do it. Well, not so fast.

Complexity is an inherent problem in fields from computer science to biology to physics. It is the problem of the next one thousand years. Nanotechnology is not like building the first bridge. While it is clearly too soon for the vision of nanotechnology to be a wholly engineering endeavor, the beginnings are upon us.

Nanoparticles like quantum dots and carbon nanotubes are the quintessential reality of nanotechnology today. They can exhibit new nanoscale phenomena in physics, chemical, or biological manifestations. Even if our theories of quantum physics, chemistry, and biology are complete, how they manifest themselves in complex situations is not. Progress in the nanoscience of mesoscopic and biological-materials systems will be the foundation of larger artificial small systems.

Table 2.1. Nanotechnology circa 2006, vision versus reality, a pragmatic definition

Vision

From the fantastic to the mundane the vision of nanotechnology began over 40 years ago and encompasses the building block view of building matter and machines, and artificial life. Programmable machines you can't see.

- *Examples:* atom-by-atom or molecule-by-molecule construction of everything from drugs to tiny robots; not necessarily nanometer-sized but also just small (e.g. cameras or surgical machines that can be swallowed); self-replicating nanomachines (or assemblers).
- *Threats:* unknown, but greatly imagined: prolonging human life indefinitely; uncontrollable, replicating nano-assemblers turning the world into grey goo.
- *Biggest questions:* Is this a joke? How will modifying or extending the human body change civilization? What about being able to "grow" anything cheaply with minimal materials?

Reality

The science and technology behind nanotechnology today predominately defined as new phenomena that emerge in the nanometer length regime in physical, chemical, and biological systems.

- *Examples:* nanoparticles that are (i) many times more reactive than microparticles, (ii) have changing properties – like colour – due to quantum confinement effects, or (iii) have unknown biological effects like increased absorption in the body; advanced materials technology; quantum "mesoscopic" physics; everything and anything.
- *Threats:* possible environmental hazard as nanoparticles can be much more reactive than larger particles and may linger longer; potential risk to human and animal health due to unknown effects on the body.
- *Biggest questions:* How do we regulate these materials? How do we classify them and determine their risk?

Nano-enabling technologies

Fabrication and measurement techniques that aid in the realization of the above two definitions but do not fall within their purview. The reliable manufacturing of devices and structures with at least two dimensions at the nanoscale.

Threats and Futures and Politics

Fear as much as hope has popularized nanotechnology, from the hysterical visions of "*grey goo*" to the quite realistic fears of human toxicity. The image of machines invisible to the naked eye wreaking havoc has certainly captured the public's imagination. But aren't self-replicating robots that take over the world bad at any scale? In truth, many of the fears of robotics and artificial intelligence that have been around for decades have just been transferred to the invisible realm of nanotechnology, adding another dimension of fear.

In any case, we already have replicating killer machines – they're called viruses, and bacteria, and humans. Redoing these with non-biological matter – almost the essence of the vision of nanotechnology – may well be more dangerous, or not, but will certainly be more difficult. We have to deal with those implications now in a plenty frightening way in biotechnology and genetic engineering. We're not sure what purpose calling what some biologists are already doing "nanotechnology" serves, except to say that it expands it to noncarbon based approaches. Some of the new phenomena that nanoscale systems exhibit can be just as profoundly harmful as useful. Particularly relevant are the chemical and biological effects.

The same reactivity that makes nanoparticles powerful in military-grade explosives can also have dramatic impact on environmental systems. And *nanotoxicity* is an even more serious concern. *Nanoparticle* transmission and accumulation through the body is largely unknown. So investment into the study of environmental and health impact is necessary. The real issue is why there has been so much funding for the societal implications of nanotechnology. The initial reason is to deal with the vision – but the fears of the vision are ridiculous at this early point in time. A reputable scientist would find it laughable. Seriously, where are the nanomachines? Stuck on a chip in a refrigerator in the basement of some university department? Funding the study of futures like this doesn't seem responsible. Biology today is much more scary.

The possible societal harm of nanotechnology pales in comparison to the prospect of an avian flu epidemic, designer babies, or proliferating suitcase nukes. Even so, the funding of bioethics is currently minimal and of questionable value. We believe the truth is more indirect. Industry is desperate not to repeat the fiascos of *genetically modified organisms* (GMO) in food – now banned in Europe - and to a lesser extent nuclear energy. What better way to "educate" the public then to fund a mass of public affairs and ethics professors to study the societal effects of nanotechnology? People tend to be more positive toward ideas when financially dependent on them.

Let's be clear, many of the fears disseminated to the public via science fiction and overzealous technology pundits or "futurists" are irrational; here education is important and worthwhile. Nor are professors and thinkers doing anything untoward. But it is interesting to note that the calls for all-out moratoriums on nanotechnology, ridiculous though they are, come exclusively from independent non-profit groups. So is the government's largesse enlightened or subtle propaganda? We can't know, but a prominent political theorist, Landon Winner, has called on the United States Congress not to create a nano-ethicist full employment act. What are important in nanotechnology are science, the progress, and the serious considerations of how to manage the environmental and social changes it creates as it develops.

When it comes down to it, saying no to nanotechnology may be like saying no to tackling the biggest problems of our time. We are faced with epic shortfalls in energy and clean water, and environmental pollution. Nanotechnology related advances in our understanding of ever-interconnected physical, material, chemical, and biological systems are our greatest hope for beating these huge problems. We must not stop science and technology.

What's in a Name?

When asked about nanotechnology in an interview for the San Francisco Chronicle in 2004 former *Intel* CEO Craig Barrett made some illustrative remarks that are worth repeating: "Nanotechnology is a buzzword that you in the press have popularized, and the government popularizes it. The formal definition of nanotechnology is anything below 100 nanometers. Every transistor that we make is below 100 nanometers. So, if you want to know the investment that we're making in nanotechnology, it's the total investment that Intel is making. ... We don't happen to be the nanotechnology that you popularize with carbon nanotubes and quantum dots and organic molecules that are going to replace CMOS transistors. Most of that is an esoteric and populist impression of what nanotechnology is. The bulk of nanotechnology is what companies like Intel and Texas Instruments and others do today."

Is Intel really the largest nanotechnology company on earth? Is everything smaller than 100 nm nanotechnology? Or is this just CEO spin? Certainly under our bipolar transition of nanotechnology above – the combination of vision and reality – Intel's latest and greatest CMOS transistor evolution is not nanotechnology. It's just the logical progression of microtechnology. First of all, companies like Intel are not taking advantage of any new nanoscale phenomena. In fact, quantum effects (like leakage via electron tunneling) are a major problem in the most advanced integrated-circuit design. Nor are they building transistors from the ground up atom-by-atom with molecules. It may be that as the biggest dimension of Intel's transistors fall below the 10 nm length scale they will have to make use of quantum size effects or switch to another paradigm – then maybe they can claim the title.

Definitions matter in contexts relevant to nanotechnology, including patenting, risk analysis, proper funding allocation, and education. Like it or not, nanotechnology is largely perceived by the public in two ways. One, as Star Trek and most science fiction conveys it: "machines you can't see" – the more fantastic vision perspective. Nanotechnology has become what we imagine it to be. From the self-cleaning walls, to the nanites of Star Trek, to the grow anything future of The Diamond Age. Second, the public sees the toxicology issue – dangerous particles in food and cosmetic products. This is exaggerated by the wide promotion of the term as marketing tool whether justified or not. The Woodrow Wilson Center has compiled a database of all products that claim to use nanotechnology. But the companies that are actually using the nanotechnology of today – nanoparticles – in situations that may be hazardous (such as in cosmetics) do not advertise this fact.

Nanotechnology has clearly caught the public's imagination, but they are missing the great stuff – the reality. There is a complete disconnect between what the public perceives and fears and what government agencies are defining as nanotechnology. Is the public seeing more than the toxicology or nano-robot issues? The last century taught us that chemicals could do truly awful things. Can nanotoxicity really be all that worse or is it just a scary name for more of the same? Meanwhile, governments and corporations around the world are funding nanotechnology for the promise of the next big thing and as a unifying theme in materials science development. Understanding how society – through the policy makers, thinkers, scientists, engineers, speakers – are defining and shaping the umbrella that is nanotechnology is our goal.

Definitions ad Infinitum

Definitions for nanotechnology abound from government program managers and institutions, corporate entities, individual researchers, and non-profits. Let us collect and analyze some of them here. Since these are the institutions that are driving the debate and/or funding the evolution of nanotechnology, this is a worthy goal.

The strongest shaper of the direction of nanotechnology is the United States federal government acting through its research funding agencies, primarily the National Science Foundation and various military equivalents (DARPA, ARDA, ARO, NRO, etc.). Mihail Roco, current head of the National Nanotechnology Initiative (NNI) and cheerleader for it even before its inception under President Clinton has given a number of definitions of nanotechnology that are often quoted. Since the NNI is shaping the future of nanotechnology by the power of federal grants, this is a good place to start. Roco's definitions are: "Nanotechnology is the creation of functional materials, devices, and systems through control of matter on the nanometer length scale, exploiting novel phenomena and properties (physical, chemical, biological) present only at that length scale."

There is an alternate definition as well: "The field of nanotechnology deals with materials and systems having these key properties: they have at least one dimension of about one to 100 nanometers, they are designed through processes that exhibit fundamental control over the physical and chemical attributes of molecular-scale structures, and they can be combined to form larger structures." And another: "Nanotechnology is the ability to understand, control, and manipulate matter at the level of individual atoms and molecules, as well as at the "supramolecular" level involving clusters of molecules. Its goal is to create materials, devices, and systems with essentially new properties and functions because of their small structure."

According to Roco, the NNI definition encourages new contributions that were not possible before:

- "novel phenomena, properties and functions at nanoscale, which are nonscalable outside of the nanometer domain;
- the ability to measure / control / manipulate matter at the nanoscale in order to change those properties and functions;
- integration along length scales, and fields of application."

Obviously Roco is firmly in the reality camp of nanotechnology, as are most of the definitions to be described. Before we analyze it further let's look at some similar definitions. The United Kingdom, which has been ahead of the game in considering the implications and risks of nanotechnology, put together a widely cited report in 2004. The Royal Society of London report defines nanotechnology as: "Nanoscience is the study of phenomena and manipulation of materials at atomic, molecular, and macromolecular scales, where properties differ significantly from those at a larger scale." This definition is often quoted outside of the United States.

Roco is seeking to ignite a new "man-on-the-moon" scale investment in science and technology in this country. He is looking at nanotechnology as a unifying theme to advance all advanced technologies, which are based largely on increased understanding of the sciences made in the past century. His definition is very biased and broad to encourage these opportunities. Business as well has a motivation to make nanotechnology widely used, at least for now, because "cool" sells.

Table 2.2. Four generations of nanotechnology according to NNI

First: Passive nanostructures. Example: coatings, nanoparticles, nanostructured metals, polymers, ceramics
Second: Active nanostructures. Example: transistors, amplifiers, targeted drugs, actuators, adaptive structures
Third: Systems of nanosystems. Example: guided assembling, 3D networking and new hierarchical architectures, robotics, evolutionary
Forth: Molecular nanosystems. Example: molecular devices 'by design', atomic design, emerging functions

Recently, the elite academic publisher *Nature (London)* spun off a new journal called *Nature Nanotechnology*. It's illustrative to find out what they consider acceptable for submission. In *Nature's* words, *Nature Nanotechnology* "is a multidisciplinary journal that publishes papers of the highest quality and significance in all areas of nanoscience and nanotechnology. The journal covers research into the design, characterization and production of structures, devices and systems that involve the manipulation and control of materials and phenomena at atomic, molecular and macromolecular scales. Both bottom-up and top-down approaches - and combinations of the two – are covered."

Meanwhile, the USA Patent and Trademark Office, trying to reorganize to deal with nanotechnology patents and simplify nanotech investment, has it's own definition: "Nanotechnology is related to research and technology development at the atomic, molecular or macromolecular levels, in the length of scale of approximately 1-100 nanometer range in at least one dimension; that provide a fundamental understanding of phenomena and materials at the nanoscale; and to create and use structures, devices and systems that have novel properties and functions because of their small and/or intermediate size."

The FDA, concerned with human exposure, takes a similar approach. "The FDA defines "nanotechnology" as research and

Table 2.3. *Nature Nanotechnology* sub-fields suitable for submission

- Nanomaterials and nanoparticles
- Carbon nanotubes and fullerenes
- Organic-inorganic nanostructures
- Structural properties
- Electronic properties and devices
- Nanomagnetism and spintronics
- Photonic structures and devices
- Quantum information
- Molecular self-assembly
- Molecular machines and nanoelectromechanical devices (NEMS)
- Surface patterning and imaging
- Nanofluidics, nanosensors and other devices
- Nanobiotechnology and nanomedicine
- Computational nanotechnology
- Nanometrology and instrumentation
- Synthesis and processing

technology or development of products regulated by the FDA that involve all of the following:

1. The existence of materials or products at the atomic, molecular or macromolecular levels, where at least one dimension that affects the functional behaviour of the drug/device product is in the length scale range of approximately 1-100 nanometers;
2. The creation and use of structures, devices and systems that have novel properties and functions because of their small size; and,
3. The ability to control or manipulate the product on the atomic scale."

We see a lot of similarities in all of the above definitions. We also see where Craig Barrett's confusion comes from. These definitions are so broad as to encompass virtually all developments in materials science and biotechnology. For example, although most of them make a point of requiring a "new" property to be considered, what is new?

It's not clear whether simply making something small (even in only one dimension) is enough to be considered nanotechnology. Consider two additional contrarian definitions. Two of the most vocal non-profits critical of unfettered nanotechnology investment are Greenpeace and ETC Group. Greenpeace's definition is: "The most common definition of nanotechnology is that of manipulation, observation and measurement at a scale of less than 100 nanometers (one nanometer is one millionth of a millimeter). However, the emergence of a multidisciplinary field called 'nanotechnology' arises from new instrumentation only recently available, and a flow of public money into a great number of techniques and relevant academic disciplines in what has been described as an 'arms race' between governments.

Nanotechnology is really a convenient label for a variety of scientific disciplines which serves as a way of getting money from Government budgets." The ETC Group definition: "Nanotechnology refers to the manipulation of living and non-living matter at the level of the nanometer (nm), one billionth of a meter. It is at this scale that quantum physics takes over from classical physics and the properties of elements change character in novel and unpredictable ways."

The ETC Group definition is focused only on the quantum nature of some nanoscale phenomena, accounting for only a part of the reality side of nanotechnology. The Greenpeace definition is very politically astute. It's interesting though that neither include in their definition anything about risk. It will prove interesting to check the longevity of these definitions a decade from now.

They are evidence of the shotgun approach the professional community is taking toward nanotechnology. There has been some backlash to this from some of the scientific community. One example is a recent *Nature* commentary by French nanoscientist Christian Joachim. He suggests defining nano as: "Nanoscience should be reserved solely for the study of a single atom or a single molecule, that is, of one entity at a time, not for groups of such entities where statistics or interactions between them come into play." This too narrow definition has good intentions. The ambiguity of the term nanotechnology sometimes attracts students with misperceptions about what they are getting into. Joachim suggests that a student might find themselves in a "physics of microelectronic devices" class instead of one on imagined nanomachines. Of course, one could argue that attracting students to take a physics class is good, misperception or not.

Perspectives from Science

Most nanotechnology definitions are cast in the language of application. This is good, since we are trying to describe a technology that does have applications now and in the future. Indeed, the language of application may be what unites all the subfields that make up nanotech and makes it a viable societal construct, though not necessarily a scientific one. Nevertheless, there are serious reasons not to lump too broad an array of advanced technologies under one banner. Education is one example. Patenting is another. Labeling inventions "nanotechnologies" sounds like a good idea for promoting interest and investment in a new and promising arena, but divorcing a patent from the specific field or principle it derives from may be devastating to the patent process.

Adequately judging prior work, quality, and worthiness may depend crucially on expertise in a specific field. There exist no experts in nanotechnology, nor will there, as society has defined the term. Here we want to comment on the "new"-ness of "*new nanoscale phenomena*", which is central to so many of the definitions of nanotech above. We also want to provide some context for the scientific sub-fields that nanotechnology is starting to claim, and show for a few examples where this is and is not worthwhile.

Condensed matter (CM) physics, formerly known as solid-state physics, is the largest subfield of physics, the modern six being astrophysics, atomic physics, biophysics, condensed matter or solid state physics, high energy physics, and nuclear physics. CM physics deals with the study of matter that is condensed, including the physics of metals, semiconductors, liquids, and emergent phenomena in many particle systems (such as superconductivity and the fractional quantum hall effect, atomic gases, as well as "soft" systems as in polymer physics). CM seeks to understand the non-classical behaviour of everything, solid and liquid, that's cold (as compared to the sun) from the scale of a few atoms to something you can hold in your hand. So its purview is quite broad and overlaps very much with materials science and physical chemistry.

In the CM community there is very little change of behaviour due to the introduction of the word nanotechnology except indirectly via funding programs. Here, emphasis on applied and interdisciplinary work has become more prevalent in recent years. CM physics has ridden (and driven) the trend in the electronics industry towards miniaturization to interesting discoveries at the nanoscale, including quantization of electrical and thermal conductance and coulomb blockade in lateral quantum dots, though often at extremely low temperatures. When CM physicists consider nanotechnology or nanoscience they usually think of one of three things: (i) quantum size effects that emerge at the nanoscale, (ii) new surface physics/energetics/mechanics in nanoscale structures due to the increased surface to volume ratio, and (iii) quantum coherent transport effects through mesoscale systems.

Largely these occur in metallic and crystallographic systems, the most common of the latter being silicon, gallium arsenide, and carbon. There is also a history of studying extremely small life - "*nanno-biology*" - but this history has been largely forgotten. The nano-bio-technology we see today in electrical engineering labs that are going "*wet*" is really a merger of hard inorganic device (CM) physics being applied to biological matter.

Cold or Hot, Quantum or Not

One could easily fill an office with books on only one topic: silicon. Probably the most economically important and well-studied material on earth, it has taken almost 100 years (starting with the discovery of quantum mechanics) to almost fully understand this simple periodic structure. (If it has taken this long to master what's basically a simpler version of sand, how long will it take to understand biological systems?) Consider an atom of silicon. It has discrete energy levels or "*shells*" where electrons can exist. As the atoms come together to form a crystal, the energy levels become energy bands, continuous as a function of the electron's position within the crystal lattice. Since silicon is a semiconductor, its optical properties are largely governed by a band-gap that separates the valence (or core) electron bands from the conduction bands, where the electrons are much more mobile. Semiconductor technology is quantum, in some sense, as it requires a quantum mechanical description of matter to explain.

The progression of a block of silicon crystal from the macroscopic scale down is a great way to understand some of the key manifestations of new nanoscale phenomena, but we could use many systems. Nanotechnology proponents often bring up new quantum behaviour that emerges at the nanoscale. There is nothing new, *per se*, about this phenomenon - being governed by well-known quantum physics - but only recently could matter be controlled well enough to reach this scale.

There are two ways in general to see quantum effects: make a material colder or make it smaller. This is because the properties of materials, from atoms to molecules to metals and crystals, are usually governed by the interaction of electrons. And the properties of electrons are governed by the energy levels where they are allowed to exist. Crudely speaking, temperature blurs these energy levels so that electrons can hop from one level to another more easily. Here is the key point: if we make a material small enough, these energy levels will move farther apart, making this hopping due to temperature much more difficult. So, in some sense, making an object very small makes it "*colder.*" Other new properties appear as well, such as a change in the optical excitation response – as this is defined by the spacing of these energy levels as well. This phenomena, which happens to take place around the nanoscale (say 10 nm for silicon) is what is often referred to as "*quantum size effects.*"

Nano in Nanoparticles

Consider *nanoparticles*, the nanotechnology of the present. These include *nanocrystals* (also called *colloidal quantum* dots in some cases) made of either metal or semiconductor as well as the carbon fullerene family, including buckyballs and *nanotubes* (the two other isotopes of carbon). *Nanowires* are classified under nanocrystals since they are solid objects as opposed to hollow nanotubes. All nanoparticles have at least two dimensions in the nanoscale regime, 1-100 nm. They have new properties that appear because of their size that would disappear in the micron regime.

These "new properties" can be different in origin and even coincidental. We categorize them broadly into three bins: (i) new quantum effects which appear due to confinement, which are often called quantum size effects; (ii) new surface physics or much increased reactivity, usually attributed to increased surface to volume ratios – this might also include structural changes from adding nanoparticles to a material (like rubber) to make it stronger; and (iii) new biological properties which are largely a coincidence of scale, as the body is not attuned to defend against nanosized particles in many situations. It would be easy to argue that nanoparticles or quantum dots are the only nanotechnology at this point in time. They certainly constitute most of the risk.

Nanoparticles have large surface-to-volume ratios making them extremely reactive. They also are able to penetrate sensitive areas of the body and accumulate there. They can be embedded in larger substrates such as polymer systems or fabrics, which may decay over time and release these nanoparticles into the environment. They are also found in sunscreens and cosmetics. Nanoparticles and (eventually) active nanomachines bridge the materials to biology gap. Evolution in biomarkers for medical imaging and cancer treatment has tracked well with nanoparticle technology. But not all quantum dots are dangerous. For example, lateral quantum dots built in quantum well structures are no more toxic than the crystal they are in (gallium arsenide or silicon, as examples).

Not Nanotechnology

Technologies are often organized by scientific principle or concept, whereas nanotech can incorporate these just because of a coincidence in scales. Take the case of photonic band gap materials, which apply the theories historically developed for electronic waves in semiconductors to photon waves (light) traveling in periodic index of refraction materials - sometimes called *meta-materials*. Visible light photonic band gap materials will likely require nanoscale manufactured devices (<100 nm 3D structures) but work as well at microwave frequencies in order 1 cm sized structures.

The physics doesn't change whether the features are 1 nm or 1 cm. So the only thing "nano" about some photonic meta-materials may be the fabrication process. Is there any point calling it nanotechnology? Would you send this patent application to be reviewed by the "*nanotechnology*" patent clerk or the physical optics patent clerk? It should be a no-brainer (the latter). In our opinion, devices or structures with only one dimension at the nanometer length scale are in general not nanotechnology. Quantum wells are a good example. To make a quantum well you sandwich a thin semiconductor layer between two bigger semiconductor layers of a different type. This allows you to trap electrons in the inner layer. Physically this is not much different from inversion layers, where one material is grown on the other and an electric field traps electrons at the interface.

These kinds of semiconductor devices are well known. Electrons in these 2D sheets exhibit many new and interesting phenomena, often attributed to the nanoscale. But, new nano phenomena in this context could just as easily be called new quantum phenomena or new many-body phenomena. Superconductivity, quantum wells and electron gasses, *giant magnetoresistance* (GMR) are all examples of quantum behaviour misappropriated to the nano-field. This also explains some of the ridiculous money expenditure estimates - in the tens of billions of dollars some of them - which have been touted as nano-investment. If you include all the investment in GMR hard drive technology, you get big numbers.

Conclusions

Much of nanotechnology is just the natural progression of scientific and technical trends began long ago. Whether it constitutes a real paradigm shift in how we approach the manipulation and utilization of nature is still an open question. We have taken a pragmatic approach by considering the social shaping of the term as it presently stands, introducing the vision and reality viewpoints of nanotechnology. From the perspective of the scientific community, the label nanotechnology doesn't seem to be doing any harm. Funding via this route has been relatively sane, perhaps better than it would have been otherwise. Certainly investment in the cross-disciplinary developments of the fundamental science in physics, chemistry, materials science, and biology has great promise. It may even be that there is some over-investment occurring. Many states and universities across the country are building nanotechnology centers without enough qualified people to fill them.

We are left with two basic questions to ponder. How should proponents - predominately government and industry - want nanotechnology to be understood by the public? And how should we define the term in the proper sense for risk assessment? In the sole context of risk, we should immediately define a very clear definition of nanotechnology: Nanotechnology, at present, is nanoparticles and nanomaterials that contain nanoparticles. Nanoparticles are defined as objects or devices with at least *two* dimensions in the nanoscale regime (typically under 10 nm) that exhibit new properties, physical, chemical, or biological, or change the properties of a bulk material, due to their size. Nanotechnology of the future will include atom-by-atom or molecule-by-molecule built active devices.

It should be evident from the rest of our analysis that this definition is incomplete as compared to society's definition of the word. (Also note that we explicitly enforce two dimensions being in the nanoscale regime.) But it has the salience of being rigorous and relevant to the questions of nanotechnology most directly impacting the public. It might be beneficial from a public policy perspective to leave the science and technical communities alone with their developments - with their new transistors and quantum dots and so on - instead redefining nanotechnology separately, in the context only of direct and novel environmental and human impact. This would have the benefit of isolating truly worrying nanotechnology-based products from the bulk of nanoscience research, which is completely innocuous. With recent calls for a moratorium on all nanotechnology by groups such as ETC, such a redefinition might have real value.

While the breadth of phenomena that lie within the nanoscale regime points to the ridiculousness of categorizing technologies based on size, the term nanotechnology has become embedded in our society, which gives it meaning explicitly. Microtechnology, as an historical counterpoint, has referred generally to a specific set of techniques and processes with little ambiguity. But what is defined as nanotechnology today may not be nanotechnology tomorrow. Taniguchi was considering precision manufacturing: separation, addition, or removal of materials at the atomic/molecular scale. Indeed, the mechanical systems that lead Taniguchi to coin the term are now called MEMS, microelectromechanical systems, and are considered by many today not to be nanotechnology at all.

3

LABORATORY METHODS

In our culture, the negotiation of novelty is commonplace. Patenting, for example, is a process to decide what counts as novel. Innovations are compared against predecessors and consequential decisions are made on the basis of similarity and difference. The same is true of the Nobel Prizes. One might even argue that all arguments can be recast as a negotiation of similarity and difference.

Nanoscience and nanotechnology are often claimed to be novel and also often claimed to not be so. The *Oxford English Dictionary* is a useful first port of call for this kind of endeavor: The first use of the word was already in 1974 but in an obscure publication, the *Proceedings of the International Conference of Production Engineers*. The second recorded use is Eric Drexler's 1986 *Engines of Creation*, and that is of course the most important locus because this book was widely read. After 1986, one can see the word spread to publications with large readerships: *The New Scientist*, the *Times Higher Education Supplement*, the *Washington Post*, the *Sunday Times*, and *Nature*.

Drexler's *Engines of Creation* is a tremendously successful book, written in an upbeat tone of voice painting a rosy future of tremendous technological ability. Drexler argued that we can now build structures on the nanoscale, meaning that we can move and combine atoms and molecules as we do with Lego™-blocks, as long as the resultant molecules are energetically stable. We can build molecules that have similar functions as the DNA-RNA-protein system found in nature, that is to say our new molecules may be engineered so as to be parts of a self-reproducing system. From this will flow new materials, new drugs, new information technologies, new human tissues, new just about everything. In the introduction to the book, Marvin Minsky, Professor at MIT (and so a credible individual in matters technical), emphasized that Drexler's vision was not fanciful but based on a thorough knowledge of the current science and technology. The vision was compelling for two reasons. (1) The tool for moving individual atoms, the scanning tunneling microscope (STM), became well-known at just this time – it received the Nobel Prize in the same year that *Engines of Creation* was published (1986). (2) The combination of molecular biology, the incipient human genome project, and the understanding of biochemical pathways made it feasible that a slightly different ensemble than the DNA-RNA-protein one could be produced and have the same kind of tremendous power as life. Much of Drexler's book thus addresses the issue of figuring out what kinds of molecules we would want to assemble given our knowledge of molecular biology and biochemical pathways, and how to ensure that the research would be beneficial. At the very same time, in the mid-1980s, the field of artificial life came into being. Nano and A-life are natural bedfellows: one predicts new forms of life created in the laboratory, the other simulates new forms of life on the computer. Both make the creation of new forms of life in the laboratory seem less fanciful.

So, much of the feasibility of the vision depended upon the feasibility of the STM's purported control over individual atoms and upon the feasibility of alternative forms of life. And the novelty of Drexler's vision traded upon the novelty of the STM and A-life. In this paper I will focus upon the novelty of the former. As ever, this was negotiated and renegotiated.

Scanning Tunneling Microscope

What is an STM and how does it work? It was described in the following way in *Scientific American*. A very fine needle is brought very close to a sample surface, for example a crystal surface whose structure is to be examined. When very close, electrons might jump across the gap from sample to tip; especially if an electrical potential is applied (*e.g.* by connecting the tip to a battery and the sample to earth). The jump across the gap is explained within quantum mechanical theory by the phenomenon of tunneling. The electrons tunnel through the vacuum despite the classical, non-quantum mechanical theory predicting that they do not have the energy to surmount the obstacle provided by the vacuum. The tunneling electrons amount to an electrical current that can be measured with great precision. Quantum theory predicts that the tunneling current is very sensitive to the distance

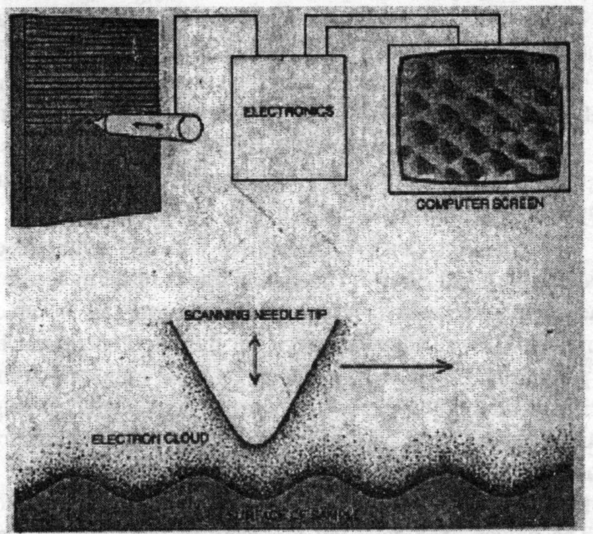

Fig. 3.1. Scanning tunnelling microscopy.

between tip and sample: proportional to the inverse of the distance squared. If one scans the tip across the surface, the distance between tip and sample will oscillate and so will the current. The correlation of tip position and current can thus be used to produce an image on the computer screen giving a rendition of the topology of the sample surface.

The STM was invented by Gerd Binnig and Heinrich Rohrer at IBM Zurich in 1981. Their very first paper was concerned with a tunneling microscope. They argued that they were able to reduce the distance between probe and surface to the dimensions of a single atom. The proof of this lay in the tunneling current measured (inversely proportional to the distance squared, a proportionality that is theoretically explainable only with the quantum mechanical notion of electrons tunneling across the vacuum between probe and surface). The main point here is that they relied on quantum mechanics. They themselves highlighted the fact of atomic resolution: "Surface microscopy using vacuum tunneling is demonstrated for the first time. Topographic pictures of surfaces on an *atomic scale* have been obtained."

Holy Grail of Atomic Resolution

In order to understand this, let us examine the significance of the term "atomic resolution" for the audiences that Binnig and Rohrer addressed. With some hyperbole one might say that atomic resolution had been the holy grail in the natural sciences for at least a hundred years. 19th-century scientists developed a language based on atoms as elementary building blocks with which all sorts of analytical and industrial chemistry was carried out. The concepts of the atom and of Mendeleev's elementary table were tremendously useful. But it was agreed that there was no direct evidence of atoms and many scientists developed a pragmatic attitude, dismissing all discussions of atoms as metaphysical – beyond measurement, beyond our ken.

In the early 20th century, much experimental evidence emerged with radioactivity and x-rays. The visible tracks made by alpha particles in cloud chambers were very powerful, and atom-talk became kosher once more. William Henry Bragg, for instance spent much of his career popularizing such talk, lecturing on BBC radio and at the Royal Institution on individual particles flying through a gas. He also spent much time developing x-rays as an analytical tool in crystallography. A broadside of x-rays will be deflected at a crystal surface, and the many deflected waves combine to produce a pattern on a photographic plate. The power of X-rays lay precisely in their atomic resolution: they yielded information on the average distances between atoms in the crystal lattice.

Many similar techniques were developed to explore surfaces, especially with the growth of the semiconductor industry in the 1950s and '60s. Scientists used light, electrons, or ions of all kinds of wavelengths or energies shooting at all kinds of angles at the surface, sometimes measuring the particles transmitted through the target, sometimes those reflected back. Knowledge of the structure of semiconductor surfaces was obviously of tremendous financial importance and so this armory of techniques became large and very sophisticated. The same techniques were used to examine metallic surfaces for which there was also tremendous industrial interest.

Novelty of the Scanning Tunneling Microscope

So, by the early 1980s, there was a large, if diffuse, social grouping of surface scientists, united by an understanding of, and a commitment to, an array of techniques yielding information about surfaces, often with atomic resolution, but always averaged over many atoms. In the following years, Binnig and Rohrer worked also to explain just what constituted the novelty of their new instrument. In the abstract of one paper (Binnig & Rohrer 1982) they referred to "unprecedented resolution in real space on an atomic scale" (real space in contrast to the conceptual "reciprocal space" used with diffraction techniques). They now explain: "The usual experimental methods to investigate surface structures (*e.g.* LEED, atom diffraction, ion channeling) are indirect in the sense that 'test models' are used to calculate the scattered intensity profile which is then compared with the one measured. In addition, these methods usually require periodic surface structures. The STM, on the other hand, gives 3d pictures of surface structures direct in real space".

Binnig and Rohrer not only advertised their new instrument to a busy but potentially interested audience, they also had to convince them that they were credible. Some scientists directly accused them of fraud and some reviewers rejected their papers. A knee-jerk reaction of many scientists was that the resolution of an individual atom was impossible, due to the uncertainty principle, a fundamental tenet of quantum mechanics. The fact that the quantum mechanical effect of tunneling was centrally involved will have given scientists the immediate association of quantum mechanics and its somewhat different laws for the atomic length scales. The uncertainty principle may be explained in the following way. If one were to determine the position of an individual atom, then one could send out light (a photon) which, if impinging upon the atom, would change direction.

The deflection of the photon would yield information about the atom's position, but unfortunately the deflection of the photon entails the slight movement also of the atom. Thus, some uncertainty will always remain about such issues as the position of individual atoms. Most scientists learning quantum physics will learn about the uncertainty principle with examples such as the one just given. Nowadays STM users will learn that the uncertainty principle does not apply for the case of atoms embedded in a solid and that the examples used to explain the uncertainty principle apply only to free atoms. In other words, while the photon might nudge the atom, the neighbouring atoms will push it back into place. But in the early 1980s, the audience will have consisted of many busy scientists whose knee-jerk reaction when hearing of atomic resolution of individual atoms was to dismiss it.

Some scientists will also have had much investment in the existing techniques and have been reluctant to accept a new one that might render their expertise obsolete. Surface scientists and crystallographers were, generally speaking, proud of their facility to think in terms of both real and reciprocal space. And so Binnig and Rohrer needed to build up their own credibility. For instance, they needed a convincing theory based on quantum mechanics explaining the tunneling process.

According to this theory it is not just a question of "feeling" the topography of the surface but rather a result of the overlap of electron orbitals of the tip and sample atoms with the greatest proximity.

The bottom line is that STM measurements require interpretation according to a theoretical model, and that it is not immediately obvious which model is the most appropriate. On top of all this, it is difficult to get the STM to work properly: proficient users will tell you that it might measure junk for hours and then suddenly yield sensible information. (This phenomenon, so the explanation goes, is due to the chance placement of an atom on the tip that gives it the required sharpness. That is to say, when scanning across the surface very closely, a surface atom might jump from the surface to the tip and sit in such a way as to jut out and give the tip the desired sharpness.)

All of this means that other scientists had plenty of reason to dismiss Binnig and Rohrer's results, and one might expect that those with a career invested in existing techniques would feel threatened by an instrument promising markedly better performance. Many surface scientists thus had both the motivation and the arguments to reject the STM. The politic reaction of Binnig and Rohrer was of the kind: 'okay guys, it's not that novel, really – relax and give us a break'.

They wrote: "we understand the STM as a complement to present microscopy rather than a competitor. For many applications, the STM is best used in combination with another microscope". And indeed everyone used the STM in conjunction with another microscope. The proficient new STM user was able to discern obvious noise from a proper measurement by comparing the result with that obtained from another tool.

The evidence yielded by the STM is mediated through quantum theoretical understanding and a profound pre-existing understanding of surfaces. (For the importance of the pre-existing understanding of surfaces.) And importantly, the novelty of the STM was negotiated: at times it was emphasized, at times downplayed. The novelty sometimes focused on the atomic resolution but it didn't have to. For example, the AFM, the sibling of the STM and much more widely used, does not yield atomic resolution. The utility of the instrument doesn't require atomic resolution. But symbolically, atomic resolution mattered greatly – comparing it to the holy grail is not too much of a hyperbole, after all. As always, a new technique becomes credible only when replicable, and it took years for an STM to be built successfully outside IBM Zurich. Other IBM labs came first and by 1985 there was a small

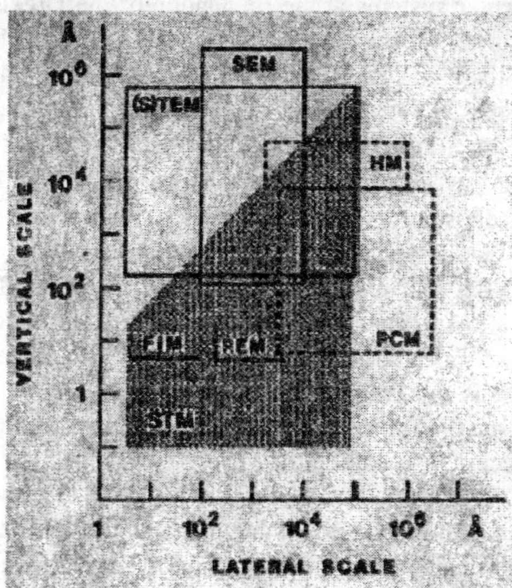

Fig. 12.2. The resolution of various microscopic technique.

community of STM users. At this point, *Scientific American* picked up the story. Binnig and Rohrer wrote the article jointly with the staff of *Scientific American*. The staff of course knew how to address a broader audience than just the surface science community, and so the language shifted importantly. The new kind of microscope enables one to "see" surfaces "atom by atom". The article also advertised the instrument's versatility: it "may extend to investigators in the fields of physics, chemistry, and biology".

The next year, 1986, was the STM's breakthrough year. Binnig and Rohrer received the Nobel Prize, and Eric Drexler published his influential *Engines of Creation* that also popularized the notion of nanotechnology. Drexler does refer to the STM, but not centrally. The manipulation of individual atoms is pretty much taken for granted, and he focuses much more on the implications of that purported ability, thus shifting the discourse towards artificial life and the creation of alternative life forms.

Hyping of the Scanning Tunneling Microscope

The story of the scanning tunneling microscope and its new siblings (collectively called *scanning probe microscopes*, or SPM) after 1986 primarily went off in the direction of immediate utility that is discussed by Cyrus Mody. One might posit a continued disconnect between the actual work done with SPMs and the LEGO™-style construction of life-like molecular systems at the foundation of the Drexlerian vision. Even the historian of science, Jed Buchwald has contributed to this disconnect by rendering an illustration of "Zippenfeld's amazing atomic etcher", purportedly for touching up the family's greeting cards. The illustration is unreferenced and Buchwald in fact made it up himself.

One event has enhanced this disconnect more than others: IBM employees' media stunt, writing IBM with individual atoms. They used "the STM at low temperatures (4K) to position individual xenon atoms on a single-crystal nickel surface with atomic precision. This capacity has allowed us to fabricate rudimentary structures of our own design, atom by atom ... the possibilities for perhaps the ultimate in device miniaturization is evident." The paper made it straight to the front page of the issue of *Nature* in which it was published. The reason for its media success was of course its relevance for the Drexlerian promise/hype. It is of methodological advantage to talk about promise/hype, to retain a Janus-faced ambiguity and not decide in advance whether nanotechnology will succeed or fail. The nature of the promise requires no further explication at this point, whereas the nature of the hype does.

First of all, the IBM experiment worked only at 4K, an extremely low temperature, and at high vacuum. One of Drexler's points was that we would only be able to assemble energetically stable molecules, and IBM's surface with patterns made by xenon atoms is not energetically stable except at these low temperatures. Furthermore, Eigler *et al.* were able to move atoms laterally on a surface, which is rather different from assembling a three-dimensional molecule – DNA, RNA, and proteins are of course not flat. In a word, there is a tremendous disconnect between moving xenon atoms on a surface at 4K, if that is what Eigler actually does, and building large complex bio-molecules LEGO™-style. Xenon, after all, is an inert gas, meaning that it prefers not to bond chemically. Nudging along an atom that skates on the surface without any propensity to engage with the substrate is comparatively easy; picking up a chemically active atom and placing it somewhere in a huge chemically active three-dimensional molecule is completely different.

Don Eigler has continued to popularize this experiment. Visitors may experience the set-up at IBM's Almaden Research Center in San Jose, California, and a virtual art gallery of STM-renditions of xenon atoms on a nickel surface has come into existence. In 1996, Charles Siebert "flew across the country to move an atom" and to write about it in the *New York Times*. That he had to fly from New York City to San Francisco indicates that we are not talking about an experiment that has proliferated greatly. Others have written or drawn other words and images with a similar set-up, but this technique is not being worked on for industrial application. Siebert ignored, and perhaps didn't even understand, the mediated nature of his movement of single atoms. All he did was to "nudge around a single atom of the element xenon, to pick it up and put it back down, to will that atom where I wanted". There's no talk of a hand using a mouse in coordination with an image on a computer screen, and still less talk of what goes into the making of that image. Eigler's program for moving the atom with the mouse even has a chirpy sound, when the atom falls into place, just as LEGO™-bricks click when slotted together. Even more than the *Scientific American* article of 1986, articles like Siebert's elide the disconnect. Obviously, promise/hype sells better than pedantic arguments.

But it is precisely the elision of the pedantic argument that is of interest here, the elision of the differences between atomic resolution, atomic manipulability, and the ability to assemble self-replicating molecular systems LEGO™-style out of individual atoms. The word nanotechnology focuses our attention on the nanoscale, the scale of atoms, and this term covers a multitude of sins. Nano is simultaneously scanning probe microscopy, Eiglerian atom nudging and Drexlerian hype.

Science Fiction

In a very interesting article, Colin Milburn has disclosed the close relationship between Drexler's arguments and the genre of science fiction. Science fiction is identified by the narratological deployment of a novum – a scientific or technological innovation extrapolated from present-day realities – that entails a change in the whole universe of the tale. "Science fiction assumes an element of transgression from contemporary scientific thought that in itself brings about the transformation of the world. It follows that nanowriting, in positing the world turned upside down by the future advent of fully functional nanomachines, thereby falls into the domain of science fiction".

Milburn shows that *Engines of Creation* is composed of a series of science-fictional vignettes, providing a veritable checklist of science-fictional cliches. He finds the same elements in the technical writings of Ralph Merkle, Markus Krummenacker, Richard Smalley, Daniel Colbert, Robert Freitas, Jr., J. Storrs Hall, "and other prophets of the nanofuture [...] Matter compilers, molecular surgeons, spaceships, space colonies, cryonics, smart utility fogs, extraterrestrial technological civilizations, and utopias abound in these papers, borrowing unabashedly from the repertoire of the twentieth-century science-fictional repertoire". Milburn even shows that Feynman's famous 1959 lecture "There is plenty of room at the bottom", which is routinely deployed as an origin myth, belongs in the same category. It too is structured in a series of science fictional vignettes and it too draws on science fiction themes of its time.

The genre is visible in official literature too, for example in that of the munificently endowed National Nanotechnology Initiative – witness its brochure *Nanotechnology: Shaping the World Atom by Atom* the main author of which seems to be Ivan Amato, an author who has also written a book extolling the virtues and promise of materials research. Drexler himself has institutionalized his bolstering role with the foundation of the Foresight Institute. There can be no doubt that the promise/hype of a Drexlerian vision has helped direct funding in a certain direction.

Technological Futures

In a book entitled, *Imagining the Future*, Joseph Corn has assembled half a dozen histories of technological promise/hype. There is for example a story about the early discourse on x-rays for therapeutical purposes (where the promise was to eliminate disease *tout court*), the electrical home (to eliminate domestic labour), or nuclear power (to eliminate war and even social strife). In the epilogue, Corn sums up the imagined technological futures as each fitting at least one of three fallacies. The first is the fallacy of total revolution; that a new technology was expected to herald tremendous change whereas the change in fact turned out to be less significant. The second fallacy is that of social continuity; whereas in fact all new technologies altered the society into which they were introduced. The third is the fallacy of the technological fix or the expectation that the new technologies would strengthen the values of old existing social patterns, whereas they turned out to introduce novel and unintended ones. The Drexlerian vision certainly heralds revolutionary change, and it talks only about the technological changes in store, ignoring and thus not expecting attendant social changes. And the latter part of *Engines of Creation* discusses how to set up an institution of oversight to ensure that the nanotechnological revolution brings only what we desire and none of the technological nightmare conjured up by, say, *Prey* (Crichton 2002).

In this sense, the Drexlerian vision seems to conform to other technological visions.

But why should technological visions have to come true? Their purpose is not to predict but to enroll. They invite other researchers to jump on the bandwagon by depicting an exciting and fruitful field. Whether or not the visions are related to science fiction does not really matter, except to the extent that they help and hinder the political project of bringing allies together. The genre of science fiction explores just what will cause broad excitement, and as such it provides a natural resource for promise/hype. But it is clearly a double-edged sword, especially because of the term "*fiction*". Much of the discourse around Drexler negotiates the proper boundary of reality and fiction. This is Milburn's main concern, along with his argument that the difficulty of maintaining that boundary contributes to a post-modern breakdown of hitherto established identities. At the same time, this negotiation is simultaneously a political dance that makes and breaks alliances.

Role of Visions

Most importantly, visions aim to increase the chances of funding. The National Nanotechnology Initiative's programmatic statement, *Shaping the World Atom by Atom*, is illustrative. The vision is science fictional in the above sense. The argument is then made that R&D funding has been geared to short-term projects with specific goals defined in a cost benefit analysis, but that the promise of nanotechnology couldn't be realized with such funding because the tremendous practical difficulties render the likelihood of short-term marketability unlikely. The role of the government, so the NNI-report, is to step in precisely in such cases as nanotechnology, where the absence of short-term returns prevent investment from private enterprise, but where the promise of long-term benefit makes it worthwhile. No vision, no funding.

In the 1990s physicists in particular have become accustomed to cuts in funding, and this may well be related to the lack of a compelling vision. The National Ignition Facility (NIF) is an interesting contemporary example. This is a facility with the aim of achieving fusion by focusing many high-energy lasers very precisely on a very small area in space, thus providing enough energy to overcome the threshold for fusion. If successful, the system would unlock even more energy than fission, and thus tremendous amounts of energy could be obtained from hydrogen atoms, far more energy than the input to start the fusion process.

The investment for the NIF is several billions of dollars and even if fusion were to be achieved, the engineering task of putting that energy to good use would only just have begun. Thus, in order to attract long-term funding, the vision has to contain much promise. Now, the promise/hype of the NIF is very similar to that for nuclear power in the 1950s. It promises the most powerful weapon ever, and thus a US monopoly, in turn ensuring global peace through deterrence. It will also provide an abundance of energy for civilian use, providing affluence to all, eventually resulting in the end of social strife. Presumably the similarity with the chiliastic nuclear vision is both a source of strength and weakness. The many political alliances that are already in place sustaining nuclear power are likely candidates for enrolment, but for the same reason the well-organized enemies of nuclear power will be enrolled just as easily. Furthermore, the similarity to an older and failed vision makes the NIF project look less than exhilarating. By contrast, the Drexlerian vision's piggy-backing on the promise/hype of fashionable molecular biology gives it sheen and luster.

Technological visions of the future are alive and well, but of course not any vision will do. And as predictions, they are bound to fail: sophisticated notions of the interaction between technological and social change would be counterproductive. The visions are intended to tie together the elements of a heterogeneous network that requires constant maintenance in order to hang together (Latour 2002, 2004). The claim of novelty is essential for technological visions: the elision of the connectedness with practices and theories of the past is as productive as is the claim that success has been shown to be possible in principle, requiring from now on merely developmental labour. The role of promise/hype in motivating researchers and funding bodies discussed here has not been the subject of much research so far. Studies that examine the role of the public sphere instead tend to focus upon the issue of consensus. The topic of nano provides plentiful material for future analysis.

Technological and Scientific Aspects

Since two decades *nanotechnology* has evolved from different scientific fields, such as physics, chemistry, molecular biology, and material science. The nanotechnology aims to study and to manipulate real-world structures with sizes ranging between a nanometer, i.e. one millionth part of a millimeter, and up to one hundred nanometers. The set of typical "*nano*"-objects includes colloidal crystals, molecules, DNA based structures, and integrated semiconductor circuits.

The first scientist who pointed out that "there is plenty of room at the bottom" was Richard Feynman in the year 1959. He envisioned scientific discoveries and new applications of miniature objects as soon as material systems can be assembled at the atomic scale. To this end, machines and imaging techniques would be necessary which can be controlled at the nanometer or subnanometer scale. Fifty years later, the scanning tunneling microscope and the atomic force microscope are ubiquitous in scientific laboratories allowing to image structures with atomic resolution. As a result, various disciplines of the nanotechnology aim towards manufacturing materials for diverse products with new functionalities.

There are two strategies for assembling nanosystems: the *top down* and the *bottom up* approach. The "top down" approach follows the development of the microelectronic industry to miniaturize integrated semiconductor circuits. Modern lithographic techniques enable to pattern nanoscale structures, such as transistor circuits, with highest precision down to several nanometers. In 1965 Gordon Moore, co-founder of Intel corporation, predicted that the number of transistors on a computer chip would double about every eighteen months. The exponential law, also known as *Moore's first law*, has described the development of integrated circuits surprisingly well for decades. As the market for information technology continues to grow, the demand for computer hardware instigates more and more sophisticated "top down" techniques to build more densely packed transistor circuits. *Moore's second law* states that the implementation of a next generation of integrated circuits at minimum cost will be exponentially more expensive as well.

Until all the constraints will finally limit the growth of the semiconductor "*top down*" industry, scientists and engineers assume that the nanotechnology will give answers to most of the technological challenges. For instance, as soon as the feature size of the semiconductor transistors reaches the level that quantum phenomena are important, different concepts for the assembly need to be considered. One possibility is the "bottom up" approach, which is based upon molecular recognition and chemical self-assembly of molecules. In combination with chemical synthesis techniques the "bottom up" approach allows assembling of macromolecular complexes with new functionalities.

Nanoscience and *nanotechnology* will definitely have a strong impact on our lives in many separate areas; e.g. information technology, material sciences, and medicines, to name only a few. The manuscript starts with a paragraph about the most recent developments of lithographic techniques in industry and nanoscience. Moreover, most of the discoveries of the nanotechnologies were initiated by the development of accurate microscopes with atomic resolution. Therefore, the manuscript also gives a short introduction to scanning microscopes. A paragraph about nanoelectronics features approaches and techniques which are supposed to produce successor technologies of the microelectronic industry. Examples include quantum computing and spintronics, two emerging fields of nanoscale electronic circuits. Alternative materials and approaches are currently being investigated for novel products consisting of nanostructures. Aspects of nanoscale materials as well as the impact of the nanotechnology on health sciences are finally described in the manuscript.

Top-down Technique

The optical "*top down*" lithography has been the backbone of the semiconductor industry since 45 years. In this technique a so-called *photoresist*—a liquid, photosensitive chemical that resists etching processes—is spin-coated onto the polished surface of a semiconductor wafer. After hardening of the photoresist, e.g. by heating, the semiconductor sample with the photoresist layer on top is exposed to light. Hereby, patterns can be defined in the photoresist. After a developing process, the exposed (or the unexposed) parts of the semiconductor wafer are bared. By follow-up processes such as etching, metallization, and oxidation the patterns can be translated to conductor paths, logical circuits, or memory cells. Since the eighties of the last century, the demise of the optical lithography has been predicted as being only a few years away. However, each time the optical lithography has reached a limitation, new techniques extended the economically useful life-time of the "*top down*" technique.

The minimum feature size of optically defined patterns depends on the wavelength of the utilized light as well as factors, which are due to e.g. the shape of the lenses and the quality of the photoresist. In 2003, the line-width of semiconductor circuits fell below one hundred nanometers; i.e. the semiconductor industry can be literally seen as being part of the nanotechnologies. For this achievement, Argon Fluoride Excimer lasers are applied with an optical wavelength of 193 nm in the deep ultraviolet. For lithography in this optical range tricks such as *optical proximity correction* and *phase shifting* were invented and successfully implemented. On one hand, the miniaturizing of semiconductor circuits is limited by economic costs for the semiconductor industry, since the implementation of new techniques to realize an ever smaller feature size results in ever increasing costs. On the other hand, physical material properties, such as the high absorption level of refractive mirrors at short optical wavelengths, give a natural limit of only a few tens of nanometers for the miniaturization process. Assuming Moore's laws, the final limit for the optical lithography is supposed to be reached in less than a decade.

At present, there are several possible "top down" successor nanotechnologies: e.g. the *extreme ultraviolet light technique* (EUV), the electron beam lithography with multicolumn processing facilities, the *focused ion beam technique* (FIP), and last but not least the ultraviolet nano-imprinting technique. The implementation of each of the above techniques implies vast technical challenges to overcome. The most promising technique is the one utilizing extreme ultraviolet light of a wavelength of only thirteen nanometers. For this technique, fabrication errors of the "optical" components need to be in the nanometer or subnanometer range. For comparison, the state-of-the-art x-ray telescopes, such as the *XXM Newton* telescope of the Zeiss AG, exhibit a granularity of the mirror surfaces of 0.4 nm. In addition to the "state of the art" optical requirements, all metrological components of the extreme ultraviolet technique need to exhibit subnanometer resolution. As a result, the costs of a stepper machine, by which photoresists in the extreme ultraviolet can be exposed, are fifty millions dollars per system; enabling a line-width of 35 nanometers.

For medium-sized businesses, the *ultraviolet nano-imprinting technique* seems to be the most promising method to fabricate nanoscale circuits. Here, nanostructures are mechanically imprinted into a photoresist. The stamp with the nanoscale patterns is made out of fused quartz; a material which is transparent for ultraviolet light. As soon as the stamp is plunged into the photoresist, a short ultraviolet light pulse causes the photoresist to polymerize according to the patterns on the stamp. In line with the optical lithography, follow-up processes allow defining nanoscale circuits in various geometries. The minimum feature size of the nano-imprinting technique is about ten nanometers. At the same time, the technique is applicable to metals and plastic materials. The costs for an industrial nano-imprinting machine are supposed to be less than one million Dollars. However, the throughput of nano-imprinting machines is much lower than for stepper machines.

Bottom-up Technique

The antipole of the "top down" approach is the so-called *bottom up technique*. Generally, "bottom up" assembly techniques seek to fabricate composite materials comprising of nanoscale objects which are spatially ordered via molecular recognition. The primary examples of the technique are *self-assembled monolayers* (SAMs) of molecules. A substrate, usually made out of metals, is immersed into a dilute solution of a surface-active organic material that adsorbs onto the surface and organizes via a self-assembly process. The result is a highly ordered and well-packed molecular monolayer. The method can be extended towards *layer-by-layer* (LBL) assembly; by which polymer *light-emitting devices* (LED) have already been fabricated. The self-assembly technique also allows positioning of single molecules in between two metal electrodes, and subsequently into an experimental circuit. By this setup, quantum mechanical transport characteristics of single molecules, such as photochromic switching behaviour, can be studied in order to build electronic devices with new functionalities.

Monolayer Techniques

There are several combinations and variations of the self-assembled monolayer techniques. It can be combined with nano-imprinting methods, the atomic force microscope, or the focused ion beam technique; allowing the fabrication of geometrical patterns of molecules with varying friction, chemical functionality, and/or topological characteristics. Since most of the processes are performed in solution, electrical fields give further possibilities to assemble charged compounds in a directed way; e.g. for creation of functionalized sensing electrodes. Electrodes modified by negatively charged gold nanoparticles, and positively charged host molecules have proven highly sensitive for the detection of e.g. adrenaline (as the guest molecule). A further very promising field is the DNA directed assembly of network materials. Here, molecular recognition reactions between single DNA strands are translated into aggregate formation of other materials such as nanoparticles. For instance, the molecular recognition of DNA molecules has been exploited to build so-called nano-tweezers. Last but not least, material networks can be built by so-called block copolymer templates. A typical application of this technique is the formation of networks of metallic nanowires; i.e. metals are vapour-deposited onto a preformed template matrix made out of copolymers. The polymer networks can be used as two- or even three-dimensional templates, while the glass transition of the polymers gives further flexibility to form such porous films.

Generally, there is a wide range of materials which can be lithographically engineered by “bottom up” techniques. The corresponding research has led to numerous sensing, electronic, optoelectronic, and photoelectronic interfaces as well as devices. Prominent examples of nanostructures, which can be assembled with "bottom up" techniques, are presented below in the paragraph entitled "*material sciences*".

Scanning Microscopes

Scanning microscopes are key inventories for the nanotechnology. On one hand they allow probing the characteristics of nanoscale objects with highest resolution. Examples include topology, material configuration, electrical, chemical, magnetic, and optical properties of the studied objects. On the other hand, scanning microscopes allow local manipulation of the nanostructures. In a seminal work in 1982, Binnig and Rohrer invented the *scanning tunneling microscope* (STM). Piezo crystals move a scanning tip across the surface of a sample, while the electric current is recorded between the tip and the sample. If the tip is located very close to the surface of the sample, the electric current comprises of tunneling electrons. To simplify matters, electrons behave as waves in the quantum world. As soon as the distance between the tip and the surface is in the order of the electron wavelength, electrons can tunnel from the tip to the surface. The quantum nature of the tunneling process ensures that the point of the tip, which is closest to the surface, contributes mainly to the current. In principle, this point can be made up of only one atom, which allows atomic resolution.

The scanning tunneling microscope is sensitive to electronic densities on the surface of the sample, which allows imaging of electronic orbits of atoms. At the same time, the scanning tunneling microscope can operate at an atmospheric pressure down to high vacuum conditions at various temperatures, which makes it a unique imaging and patterning tool for the nanosciences. E.g. scanning tunneling microscopes are utilized to pattern nanostructures by moving single atoms across surfaces, while the corresponding change of the quantum mechanical configuration can be recorded *in situ*.

The atomic force microscope operates similarly to the scanning tunneling microscope. Here, the force between the scanning tip and the sample surface is extracted by measuring the deflection of the tip towards the sample. Again the atomic force microscope can be utilized as an instrument to image and to manipulate structures on the nanometer scale. Most importantly, the atomic force microscope is a unique tool to measure forces between two nanoobjects with a resolution of only a few PikoNewtons (a millionth of a millionth Newton). By functionalizing the scanning tip chemically, a single molecule can be utilized for sensing applications; e.g. the binding forces between ligands and receptors can be determined for biological and health applications. Furthermore, magnetic tips can be exploited to image the magnetic topology of magnetic nanostructures which are used for information science purposes. The atomic force microscope can further be used as a miniature groove, e.g. which cuts through nanoscale electronic circuits.

Generally, the resolution of an optical microscope is limited to approximately the wavelength of photons; typically several hundreds of nanometers. In order to resolve smaller structures there are two strategies, either using particles with a shorter wavelength, e.g. electrons, or utilizing a technique called *scanning near-field microscopy* (SNOM). The former possibility gives rise to the *scanning electron microscope* (SEM). Modern scanning electron microscopes can resolve topological variations and chemical configurations of a sample at the subnanometer length scale.

To this end, a sample is located in a high-vacuum chamber, and a beam of electrons is focused onto the surface of a sample. The reflected electrons are then detected as a function of the position of the initial beam. In order to study the optical properties of nanoscale structures, which are substantially smaller than the photon wavelength, the *scanning near-field microscopy* (SNOM) is an excellent method. Here, glass fibers with apertures of only hundred nanometers or less are utilized. The size of the aperture defines the spatial resolution of the technique, while the intensity of the gathered light is recorded as a function of the position of the glass fiber.

NANOELECTRONICS

The field of *nanoelectronics* aims towards developing devices comprising of nanostructures with new overall functionalities for the information technology. By exploiting electronic, optical, and magnetic properties of nanostructures, integrated circuits and memory devices are developed for faster and more reliable information processing schemes at lower heat dissipation and with a greater portability of future computers. Since the seventies of the last century, so-called quantum wells in semiconductor heterostructures have been subject of intensive scientific research. Heterostructures are semiconductor crystals with composite monolayers of different atoms. Most importantly, heterostructures can be grown in which electrons are confined solely to the plane of the *quantum well*. These investigations on nanoscale monolayers of composite semiconductors yielded to applications such as the CD drives, the laser printers, and the input-amplifiers of cell phones; investigations which were honored with the Nobel Prize in 2000.

Since approximately fifteen years, so-called nanocrystals or colloidal quantum dots are promising compounds of a successor nanoelectronic technology. Generally, the colloidal quantum dots are semiconductor crystals in which an electronic quantum state is localized within a tiny volume. Most importantly, the optical and the electronic properties of quantum dots are defined by the quantum mechanical characteristics of the electron state. A nanoelectronic device, which is very similar to a quantum dot, is the so-called *single electron transistor* (SET). The transistor relies on the switching of a single electron. The advantage of such devices would be the combination of minimum size with a minimum power dissipation, both of which become very important issues for densely packed logical circuits.

At the same time, optical processing schemes are expected to become important parts of the future information technology. Only recently Raman lasers, optical waveguides, and fast optical switches have been realized on silicon chips. The realization of an all-silicon based optical processing scheme is an important step to combine fiber optics with the present information technology at low cost. So far optical signals in fibers can be switched by micro-mechanical elements, and/or optical information is translated into electronic signals e.g. by photo diodes. Subsequently, the information is processed via electronic transistor elements and in turn, it is transferred again to optical information by a laser. This very inefficient scheme is expected to be replaced by the silicon-based technology mentioned above. In future devices, further nanostructures, such as *photonic crystals*, are expected to play an important role to control light within a semiconductor crystal directly. In "photonic crystals" the optical constants of a material are modulated in three dimensions by nanostructuring the (semiconductor) host crystal.

As a result, only photons with a defined wave vector and energy are allowed within the periodic structure. The incorporation of nanoscale defects into a photonic crystal further allows the deflection of certain optical wave modes within the photonic crystal. Moreover, photonic crystals can be designed, in which well-defined optical modes can interact with nanostructures. Recently it was demonstrated that by positioning a quantum dot at the center of an artificially designed optical mode volume, the electron states of the quantum dot strongly interact with the electromagnetic fields of photons. Latter experiments give rise to the feasibility of an information processing schemes purely based on quantum mechanics.

In principle, the atomic *granularity* defines a natural limit to the miniaturization of electronic circuits. The electron orbits of atoms are the smallest building blocks of a possible "nano-computer". In other words, a conductor path can not be narrower than a single atom. Therefore, it is reasonable to ask whether a computer, which exploits the quantum nature of the electron states, is technically feasible. The quantum mechanical tunneling effect is one of the more obvious effects which will arise as soon as the dimensions of the electronic circuits fall below a certain value. Since electrons are waves, they may tunnel from one conductor path to another if the paths are located in a distance which compares to the wavelength of the electrons.

Typical electron wavelengths in semiconductor crystals are in the range of several nanometers. Albeit the resulting leakage currents are small, the total effect can be significant for millions of densely packed transistors on a chip. On one hand, tunneling electrons give rise to additional heat dissipation. On the other hand, dislocated charges have to be considered as logical errors and/or they influence the general functionality of the integrated circuit via the Coulomb interaction; e.g. via *cross-capacities* between adjacent conductor paths. Furthermore, an effect called *electro migration* can derogate conductor paths made out of e.g. Aluminum; i.e. the electron current *flushes away* single metal atoms.

The quantum mechanical nature of electrons comprises properties and corresponding technical possibilities which go beyond the functionality of a classical computer. In principle, the classical desk top computer is based upon semiconductor physics with foundations in quantum mechanics as well. However, the functionality of the classical computer can be simply described by "classical" physics. The quantum computer aims towards exploiting the physical laws of quantum mechanics to make certain calculations more efficient. As far as the quantum computer is concerned, the equivalent to the binary digit (BIT) of the classical computer is the so-called *quantum bit* or *qu-bit*. In quantum mechanics, particles such as electrons are described by a set of independent parameters, which are called *quantum numbers*. For instance, the magnetic moment of an electron, the so-called *spin*, is a possible quantum number, which can be mathematically represented by a plus or a minus one half.

A quantum bit consists out of the sum of independent quantum numbers, while continuous prefactors define the weight of a specific quantum number. The superposition of quantum numbers also includes so-called entangled states. Latter are physical states of a quantum mechanical particle, where one quantum number depends logically on another in an intrinsic way. The entanglement is unknown to classical physics (and the most intuitive way to accept its reality is via mathematics). Most importantly, the superposition and the entanglement of quantum bits give rise to fast and parallel information processing schemes at a minimum heat dissipation.

A further discipline of nanoelectronics is the field of spintronics, where both charge and spin degrees of freedom of an electron are exploited to realize new electronic devices. As mentioned above, the "spin" describes the magnetic moment of an electron. It is classified by two values, which are usually referred to as "up" and "down" (the two expressions depict the fact that the magnetic moment of an electron orientates in the directions up or down with respect to an external magnetic field). The magnetic interaction of an electron spin with the crystal environment is much lower than the Coulomb interaction felt by electron charges. Therefore, heat dissipation effects are predicted to be significantly smaller for "spintronic" circuits than for "electronic" circuits. However, there are ubiquitous spin relaxation mechanisms in semiconductor crystals, by which the orientation of the spin (and thus the corresponding classical information) is lost within several hundreds of nanoseconds.

The precise fabrication of nanoscale crystalline layers has also given rise to the development of magnetic layer systems, the resistance of which alters significantly when small magnetic fields are applied. Latter effect is known as the *giant magnetoresistance* (GMR). The technology has produced the prime example of a spintronic device: the so-called *magnetic random access memory* (M-RAM). The memory concept was originally proposed by IBM, and it is now introduced into the marked of information storage technology. In this technology, the information is stored by magnetic domains, while the orientation of the domains is read out via the *giant magnetoresistance*.

In principle, the M-RAM technology combines all advantages of the present memory technologies: it is a fast, nonvolatile information storage technology, and it allows information storage at high density and minimum heat dissipation. The magneto-mechanical hard drive, present in most of the current computers, provides also a medium for high density, nonvolatile information storage. However, it is comparably slow. Hard drives, based upon a concept called *dynamic random access memory* (D-RAM), are fast, however, they need to "refresh" the information iteratively by current pulses. The "flash" memories, which are utilized in MP3-players, cell phones, and digital cameras, are quite slow and they can only be used approximately one million times. Therefore, M-RAM devices seem to be promising candidates for a "*nanoelectronic*" successor technology.

Last but not least, nanomaterial sciences give further possibilities for new architectures of integrated circuits and memory devices. For instance, a memory concept, usually referred to as *phase change random access memory* (phase change RAM), utilizes the phase transition between "*crystalline*" and "*amorphous*" states of materials to encode information. The phase transition is triggered by electrical impulses, while the resistance difference between the crystalline and the amorphous state allows identifying the binary information.

The crystalline state is re-initialized, if the material is homogeneously melted by the application of long electrical current. In principle, the method allows nonvolatile information storage at highest densities; i.e. a Terabit on the area of a stamp. A corresponding method is already implemented in *digital versatile disks* (DVDs) made out of polymers, where a short laser pulse melts small areas of the disk; i.e. the spots undergo a phase transition from crystalline to amorphous. Generally, for memory device with highest information densities, material sciences will play an important part in nanoelectronic engineering.

Material Sciences

In recent years, the material science and technology have explored a variety of new materials in which nanoscale components give rise to better properties of well known products. To this end, optical, electric, thermal, mechanical, and chemical characteristics of nanostructures are exploited to enhance certain properties of the composite systems. Examples of "nano"-products include paints, inks, cosmetics, lubricants, grinding pastes, ceramics, luminescent materials, glues, and protective lacquers; to name only a few. The new materials are envisioned to fulfill tasks, such as the ability to decompose pollutants at higher rates or to converse light into current more efficiently. For such and more complex tasks novel materials are based on several nanoscale components whose spatial organization is engineered at the molecular level. The macroscopic behaviour arises from the combination of the novel properties of the individual building blocks and their mutual interaction.

Materials with nanoscale pores, such as zeolites, are utilized for catalyzers in chemical processes or as ten-sides in cleaning detergents. Metallic alloys show a certain degree of mechanical memory. New terms such as nanotubes, nanowires and quantum dots are now common jargon of scientific publications. These objects are among the smallest, man-made units that display physical and chemical properties which make them promising candidates as fundamental building blocks of novel transistors.

Quantum dots are the ultimate example of a solid, in which all dimensions shrink down to a few nanometers. The electronic and optical properties of quantum dots are a consequence of their dimensions. At the same time, colloidal quantum dots can be synthesized from a wide range of materials. The dimension of these particles makes them ideal candidates for the nano-engineering of surfaces and the fabrication of functional nanostructures. Moreover, semiconductor quantum dots are probably the most studied nanoscale systems at the moment.

The surface of the lotus flower is *hydrophobic* due to nanoscale structures. In a biomimetic ansatz, these properties have been transferred to paints which are strongly "*soil resisting*". Investigations of the nanoscale architecture of shells and bones gave rise to new developments in the field of layered materials which are extremely robust. Carbon nanotubes have the topology of a graphite sheet which is rolled up to form a tube. These wires with diameters of only a few nanometers and a controlled chemical make-up show intriguing electrical properties and extremely high strength and stiffness. Due to their properties, carbon nanotubes are proposed to be utilized e.g. as field emitters in flat screens. At the same time, carbon nanotubes exhibit a unique ratio of inner volume to surface, which makes them ideal candidates for extremely light hydrogen storage devices. Moreover, compounds made out of carbon fibers are light and robust materials which are used in a broad range of application.

An important technique of the material sciences is the sol/gel technique, by which nanostructures are formed and controlled by the application of so-called *colloids*. Generally, colloids are systems in which droplets of one substance are formed and in turn, resolved within a second substance. Daily life examples of colloids are the "sauce bearnaise", in which droplets of vinegar are resolved in butter. Other examples are cosmetic cremes and paints. In material sciences colloids are exploited to form nanoscale structures with chemically controlled composition and/or nanoscale structures which tend to build networks as soon as the solvent is removed.

By utilizing the sol/gel technique, protective lacquers of a variety of substances have been produced. E.g. silicon nanostructures can be assembled in the liquid phase, and in turn, they are sprayed onto the surface of arbitrary surfaces. After the solvent has vapourized, appropriate heating gives rise to the formation of a ceramic cover layer which is extremely thin and hard at the same time. Due to the large surface to volume ratio of nanostructures, such ceramics can be formed at rather low temperatures. The sol/gel technique can be even utilized to fabricate optical components such as glass fibers and frequency doublers.

A whole class of nanomaterials is defined by so-called *aerogels*, which are highly porous materials consisting mainly out of "air". Again, a daily life example of such a material can be found in the French cuisine. The "Baizer" consists of white egg in which thousands of microscale air bubbles are incorporated. The air bubbles in "*Baizer*" have a size of several microns. As a result, visible light is scattered at the air inclusions and "Baizer" appears as a white substance, while white egg is transparent to visible light. If the enclosed air bubbles have a submicron diameter, aerogels appear to be transparent.

At the same time aerogels are very good heat insulators, since heat can not circulate in the air bubbles. Hereby, window glasses with remarkable insulator properties can be fabricated. Aerogels with nanometer inclusions can be produced by the sol/gel technique; i.e. spheres are formed within a colloid by utilizing networking nanostructures. A gel is formed as soon as the solvent is removed while the volume of the sol is kept constant.

The large inner surface of an aerogel gives rise to several applications in the energy sector. Lithium batteries with enhanced storage characteristics have been built. Electrical capacitors of up to 2500 Farads have been fabricated by aerogels. Last but not least, better fuel cells can be envisioned by this technique.

Health Sciences

There are several applications of *nanotechnologies* which are already implemented in commercially available cosmetic products. For instance, nanoscale spheres, made out of apatites and proteins, are implemented into toothpastes in order to enable *biomineralization* of the teeth, because the corresponding nanospheres consist out of the same material as teeth. Other well-known examples are sunscreens which contain colloids of zinc oxide. Such a sunscreen gives an excellent protection against ultraviolet radiation, which is simply reflected by the nanoscale structures, while the sunscreen appears transparent to human eyes. In addition, aerogels can be utilized for a very efficient drug delivery. Nanoscale capsules with functionalized molecular linkers can be utilized to deliver drugs to metastases via molecular recognition.

Recently, a method called *magnetic fluid hyperthermia* was introduced to oncology, which is based on heating tissues for therapeutic purposes. Generally, tumour cells exhibit a hypersensibility to heating (*hyperthermia*). At the same time, surfaces of magnetic nanoparticles can be functionalized in a way that they accumulate only/mainly in tumour tissue. Therefore, if an alternating magnetic field is applied, the dissipation losses, induced due the movement of the magnetic colloids, heat up and, in turn, destroy the tumour cells. By solving the biochemical and physiological specificity problem, cancer-specific hyperthermia protocols have been developed.

A well-known vision of scientists working in the field of nanotechnology is the "lab on a chip". The chips would have a size of roughly a square centimeter, while millions of nanoscale instruments would analyze droplets of liquids for biological, medical or forensic purposes. One possibility to move the liquids across the surface of the chip is given by so-called *surface acoustic waves* (SAW). The latter technology has recently been realized for commercially available products.

To conclude, nanoscience and nanotechnology have a great potential to solve several challenges of the modern society. Moreover, there exists a great economic expectation for growth related to nanoscale products. At present, there are about 450 companies alone in Germany, which manufacture products related to nanomaterials. To this end, the manuscript intends to give an introduction to some of the nano-technologies which will become important to the daily life in the near future; in particular these are the information technology, material sciences, and health sciences.

4

APOCALYPTIC NANOTECHNOLOGY

Nanotechnology is a rapidly developing field of technology that seems to have the potential of great upsides and excessive downsides. Thus far, there has been a strong tendency in the debate on nanotechnology to focus on either the first or the latter. Accordingly, assessments of nanotechnology tend to radically diverge.

On the one hand, optimistic visionaries promise truly utopian states of affairs, e.g. solving the problem of hunger in the world or expanding our maximum life span. On the other hand, pessimistic thinkers draw worst-case scenarios in which, for instance, nanotechnology has exceptionally disruptive effects on societies or swarms of nanoassemblers devour the whole biosphere.

The utopian views follow from one-sidedly focusing on the potential benefits of nanotechnology, whereas the apocalyptic perspective results from giving exclusive attention to worst-case scenarios. These radically diverging assessments that have thus far dominated the debate on nanotechnology seem to a lack common ground.

This situation holds the risk of conflicts and unwanted backlashes. Hence, the present state of the debate on nanotechnology calls for the development of a more balanced ethical view. This contribution will first briefly describe the field of nanotechnology. Next, the present state of the ethical debate of this field will be described as overshadowed by utopian dreams and *apocalyptic nightmares*.

Finally, a method will be introduced to develop more balanced ethical views on nanotechnology. Thus, the focus of this chapter is on the *methodology* and not on *normative analysis*.

NANOTECHNOLOGY

Terminology and Basic Idea

Nanotechnology is a rapidly developing new field of research. In the literature, both a fairly broad as well as a rather narrow concept of nanotechnology are employed. The first signifies any technology smaller than microtechnology. In contrast, the latter stands for the technology to program and manipulate matter with molecular precision and to scale it to three-dimensional products of arbitrary size.

The basic idea of nanotechnology, used in the narrow sense of the word, is to employ individual atoms and molecules to construct functional structures. In his lecture "There is plenty of room at the bottom", at the annual meeting of the *American Physical Society* at the *California Institute of Technology*, the famous physician Richard Feynman already speculated on radical forms of miniaturization. His reflections, however, were not taken very seriously at first. Both practically and theoretically, significant progress in the field of nanotechnology started only in the eighties.

Practical Development

A significant practical development was the fast progress in microscopy. At the beginning of the eighties, Gerd Binnig and Heinrich Rohrer, working at the IBM research laboratory in Zurich, developed *scanning tunneling microscopy*. This new technique can provide an image of the atomic arrangement of a metal or a semiconductor surface. Thus, using this new technique Binnig and Rohrer could, for the first time ever, "*map*" the arrangement of individual atoms of metals and semiconductors. In 1986 Binnig and Rohrer received the Nobel Prize for their achievements. The *atomic force microscope* represents a further development in microscopy. It enables images of materials inaccessible to the scanning tunneling microscope, for instance insulators, organic materials, biological macromolecules, polymers, ceramics and glasses.

Finally, it was discovered that scanning tunneling microscopes could also be used to manipulate nanoscale objects. For instance, in 1990, Eigler and Schweizer, two IBM scientists, were able to shape the three initials of the logo of their employer from 35 individual xenon atoms. It is expected that further developments in microscopy will be highly significant for nanotechnology.

Theoretical Development

A significant theoretical development in the eighties was the research done by Eric Drexler (1981 & 1986). His work provided a theoretical basis for the field. Besides, it brought *nanotechnology* to the attention of a broader public. Drexler argues that the laws of physics do not forbid the possibility of pre-programmed maneuvering and goal directed management of individual molecules. In addition, he elaborates on the future development of technical means and methods to arrange matter at the nanoscale. Finally, Drexler also discusses the various fields of application for future nanomachines. In discussing future nanotechnological manufacturing, Drexler comes up with the idea of the assembler: "a molecular machine that can be programmed to build virtually any molecular structure or device from simpler chemical building blocks". This nanoscale construction device can position molecules in every which way, thereby facilitating, for example, chemical reactions.

Through precise sequences of manipulations, a computer-steered assembler could thus—molecule by molecule—assemble any chemically stabile structure that it has been programmed to construct. According to Drexler, the development of universally applicable assemblers is essential for the further development of nanotechnology. Assemblers could obviously also be programmed to replicate themselves. From a commercial point of view, this would have the interesting advantage of being able to manufacture products in a feasible time frame. After all, if accomplished by only one assembler, constructing a car in a molecule-by-molecule way could take ages. If, on the other hand, millions or billions of assemblers could work together, things would look far more optimistic.

State of the Field at Present

In the meantime, the field of nanotechnology has developed considerably. Current research is not exclusively focused on achieving assemblers. Instead, research is directed towards the production of a wide array of different nanoscale structures. The fabrication techniques of these structures can be divided into two approaches: "top-down" and "bottom-up".

The *top-down techniques* that are used to manufacture nanoscale structures are mostly extensions of methods already employed in small-scale assembly at the micron scale, for example, *photolithography*. By further *miniaturization*, the *nanodimension* is entered. In this way, further miniaturization of microelectronics could result in nanoelectronics.

Bottom-up fabrication methods for manufacture are studied within synthetic chemistry, which is, almost by definition, the science of producing nanoscale structures. They are also inspired by phenomena such as crystal growth and self-assembly. In a certain way, many bottom-up methods try to imitate regularly occurring processes in nature. Living nature, for example, constantly shapes complex macroscopic structures from individual biomolecular elements.

It is widely expected that, if nanotechnology were to be further developed, a whole range of products could be produced more rapidly, cheaper and better. After all, manufacture would be done in a molecule-by-molecule way, meaning that all features of the product and all aspects of production could be programmed and would be under control in ways that are unknown up to now. Conformingly, many countries have set up programs to financially support further development of nanotechnology. Moreover, investing in the further development of specific nanotechnological projects is also rapidly getting more interesting for private companies.

Utopian Dreams and Apocalyptic Nightmares

The present state of ethical debate about nanotechnology seems to be dominated by utopian dreams and apocalyptic nightmares.

Utopian Dreams

During the 17th century, the idea was developed that *utopian ideals*, such as control of the natural environment, a perfect society, life without disease and pain, prolongation of life as well as enhancement of man and his characteristics could be achieved through the further development of science. All that was needed for that would be to organize science in a correct way and to work with effective methods. This modern idea became a source of great enthusiasm. Many contemporary views on nanotechnology can be placed in this tradition of scientific utopianism, which started in the 17th century.

Over and over again, it is argued that, if only nanotechnology were to be fully developed, a large part of the world's current problems would be solved and a whole array of ideals would be achieved. A few examples of the utopian perspectives of nanotechnology are given. To begin with, molecule-by-molecule manufacturing would be self-sufficient and dirt free. After all, molecular manufacturing techniques would not result in any chemical pollution whatsoever. Leftover molecules would be recycled. What is more, molecular manufacturing would enable environmental restoration at the molecular level. Unwanted chemicals could be detected and inactivated. Hence, we will be able to reverse existing environmental degradation. Next, molecule-by-molecule manufacturing could create unprece-dented objects and materials. For example, new nanotechnologically manufactured strong lightweight materials will enable easier access to space and space resources. *Nanoelectronics* could come up with computer chips that would be billions of times faster as a result of the smaller components.

Using *molecular manufacturing techniques*, we will be able to produce inexpensive high-quality products. For example, storage batteries, processors, personal computers, lap tops, cell phones and display devices could become strikingly inexpensive.

Molecular manufacturing could also be used to fabricate food rather than growing it. After all, food is simply a combination of molecules in certain configurations. Hence, the problem of hunger could be effectively solved by efficient molecule-by-molecule mass production of food. Finally, it is also in medicine that nanotechnology is said to work miracles. There are numerous astounding promises in this context. Molecular manufacturing will provide low-priced and superior equipment for medical research and health care. These improved tools would be available far and wide.

Medical nanomachines will be programmed to travel through our bloodstream to clean out fatty deposits. Hence, they would reduce the probability of cardio-vascular diseases. Medical diagnosis and drug-delivery will be transformed. Moreover, preventive medicine will be greatly improved by having nano-robots within our bodies that could provide a defense against invading viruses. Thus, nanotechnology has been hailed as the solution to many medical problems. It is even expected that it will contribute to the enhancement of man. Not only will it be possible to overcome contemporary diseases, pain and other unpleasant bodily symptoms.

Over and above, nanotechnology will enable us to enhance all our human capabilities and properties. With regard to the enhancement of the human body, it is expected that nanotechnology will enable the construction of stronger and enhanced tissues and organs. For instance, cells specific to certain tissues or organs could be reconstructed and be made immune against all known pathogens, thereby making our present immune system obsolete.

The *cryonics community* has also enthusiastically embraced nanotechnology. Cryonics means freezing people who have been declared legally dead and waiting until technology is advanced enough to reverse cause of death as well as the freezing damage. Here, nanotechnology is expected to produce real miracles in reversing all the adverse configurations of molecules in the frozen organism after it has been thawed out. Also, nanotechnology would enable an almost infinite improvement of our mental capacities. It would, for example, be possible to enhance our memory as well as all our data processing capacities. However, with regard to enhancing the human mind the scenario of '*uploading*' is the *non plus ultra*. Uploading involves transferring the contents of the human brain to a computer. Special nanomachines would scan the brain atom-by-atom. Next, the neural networks of the brain would be implemented on a computer.

Nanotechnology is finally also expected to lead to social advances. Freitas (1998b) thinks that the huge achievements of nanotechnology, especially nanomedicine, will make people more content and peaceful. It will be a great deal easier to live together in ideal harmony with perfect bodies and flawlessly functioning brains.

Apocalyptic Nightmares

Besides utopian outlooks, catastrophic scenarios have also heavily influenced the debate on nanotechnology. Severe disruption of many different aspects of society and politics is one of the nightmare scenarios that have been sketched in connection with the further development of nanotechnology. For example, rapid developments in molecular manufacturing and the concomitant inexpensive manufacturing could cause severe economic disruption. The economic upheaval could involve the sudden abundance of low-priced products, rapidly changing employment patterns (e.g. unexpected redundancy of a variety of jobs) and the problem of copying of designs.

Moreover, molecular manufacturing might also invite premeditated misuse in warfare or terrorism. First, all kinds of conventional weapons could be constructed more rapidly. Next, new *nanoweaponry* and could be made in huge numbers, low-priced, extremely powerful. Hence, traditional arms control would be far more difficult and competing states could enter a troublesome and unsound arms race that would be extremely difficult to end.

Also, infinitesimally small surveillance devices such as *nanoscale* tracking devices, *nanosensors*, *nanocameras* and *nanomicrophones* could enable dictatorial observation and control of subjects in a way that is totally unprecedented. Nanotechnology would enable total surveillance of entire civilian populations without them even noticing it.

Nanotechnology could also have the potential to cause extensive environmental damage. For example, destructive nanomachines might enter the food chain thereby disturbing entire ecological systems or nanomaterials such as *nanoparticles* could escape into the air and turn out to pose asbestos-like health treats. Some of the most serious risks of nanotechnology have been brought to the attention of the public by Bill Joy, co-founder and scientific leader of *Sun Microsystems*. Joy is especially worried about the research with regard to assemblers. After all, these nanomachines will have the worrisome capacity of self-replication. Without this kind of assemblers it is hardly imaginable how molecular manufacturing could ever become practically feasible.

After all, without the ability of self-replication, all assemblers needed for nanotechnological production would have to be built one by one, which would definitely be too expensive. In this case, however, practical feasibility and commercial viability would involve grave dangers. Technical faults, for example problems with the software of the onboard computer of an assembler, could cause unbridled self-replication. In that case, since the newly produced assemblers would also start replicating themselves, the total number of assemblers would grow exponentially. If these uncontrolled assemblers used a wide variety of raw materials as resources for self-replication, they could devour the whole biosphere in an amazingly short while. The biosphere would, so to speak, be transformed into gray goo—hence, the terminology of *gray-goo* scenario.

Another danger has to do with the fact that many private companies will try to develop and produce assemblers. After all, the perspective to put them on to the market will seem lucrative for many. Hence, it will be difficult for central governments to retain control over the development of assemblers. Therefore, there will always be the danger of abuse of assemblers by criminals and terrorists, for example, to develop weapons of mass destruction. For example, one could construct special nanomachines to destroy computer systems or certain resources of the opponent. Of course, nanomachines could also be designed to attack the enemy directly. This apocalyptic scenario, in which *destructive nanomachines* are being used for warfare or terrorist purposes, has also been called the *black-goo*-scenario.

Toward a More Balanced Ethical View

The dominance of the drastic opposition of utopian dreams and apocalyptic nightmares in the debate on the future perspectives of nanotechnology holds the risk of undesirable conflicts and unnecessary backlashes. Hence, the present state of debate on nanotechnology calls for the development of more balanced ethical views. In response to this important challenge, a six-step method is here presented. With the help of this method, a fine-grained and rational assessment can be made as to the ethical desirability of further developing research in a specific field of nanotechnology.

Specific Field of Nanotechnology

Up to now, broad and sweeping statements about nanotechnology as such have dominated the debate. However, for an ethical analysis to be sound and discriminating, it should be focused more specifically on a particular field of nanotechnological research, instead of generalizing in an all-encompassing way. Thus, more detailed and better-informed ethical research is needed. Nanotechnology is by no means one single effort. Rather, it is a complex of countless different projects with a huge variety of goals. Hence, different fields of nanotechnological research can be distinguished that will not necessarily demonstrate identical or even similar ethical aspects, for example: (i) materials and manufacturing, (ii) nanoelectronics and computer technology, (iii) medicine and health, (iv) aeronautics and space exploration, (v) environment and energy, (vi) biotechnology and agriculture and (vii) security. Evidently, the ethical assessments of developments in these fields are likely to differ as the objectives and the ethical problems encountered will be different.

Objectives of Specific Field Nanotechnology

Having specified the nanotechnological field to ethically be assessed, the next step is to focus on the objectives that the research in that field aims to achieve. Ideally, we should first discuss our needs, fundamental purposes and social ends in order to move on to make choices about ways of achieving these goals, for example by further developing certain fields of technological research. However, this is not always the case. What is more, technology often seems to develop in a seemingly autonomous way. However, it remains imperative to reflect about the goals that we try to achieve in developing certain technologies. Technology development should be directed towards a good or an end. Otherwise, technology is developed for its own sake, isolated from any human good. This would obviously be undesirable. After all, science and technology should serve man and not the other way round.

Objectives Ethics

After having specified the nanotechnological field to be assessed as well as having detected the objectives of that field, it should be asked whether these objectives are ethically desirable. After all, if the objectives that are aimed for with the further development of a certain research area are ethically undesirable, it does not make sense to continue this research from an ethical point of view.

Development of the Field

If the objectives are ethically desirable but not achievable or very unlikely to be achieved by advances in a specified field of research, it seems pointless to push the research forward. Of course, it is not possible to predict advances in science or technology in detail and with certainty. Nevertheless, in many cases it will be possible to give a broad assessment of the probability of achieving certain objectives in a specified field of research on the basis of the corpus of existing scientific literature.

Ethical Problems

The next step involves determining the ethical problems that are connected with the further developments in a specified nanotechnological field. *Ethical problems* connected with further development of nanoresearch are not necessarily the same in different fields of nanotechnological inquiry. For example, present research on nanotechnologically manufactured coatings of prostheses to improve their biocompatibility can hardly be said to pose the same ethical questions as, for instance, projects focused on producing perfect mosquito nets to reduce the problem of infectious disease or research on new nanochips that can read individual genomes in a feasible timeframe. Thus far, unfortunately, ethical problems have been mostly discussed on a very general level. As a rule, problems have not been linked to specific fields of nanotechnological enquiry. In order to give an idea of the ethical problems that have been discussed in debate on nanotechnology up to now, the most important problems are here presented without pretending to give a complete sketch.

Risks

Up to now, the ethical debate on nanotechnology has been dominated by discussions about risks. A few examples are risks of disruption of the basis of economies, environmental damage, an unstable arms race, the gray goo scenario, the black goo scenario etc. Against this backdrop, the environmental pressure group ETC has proposed that governments worldwide proclaim an instantaneous moratorium on commercial fabrication of new nanomaterials. Moreover, they want to start a transparent and global process of assessment of the various implications (for example for the environment, society and health) of the technology. Other less radical authors have called for an early and honest assessment of all the risks and unintended consequences of nanotechnology.

Equity

From a global justice point of view, the risk that the rise of nanotechnology will only be advantageous for rich countries, leaving developing countries behind, should be avoided. Therefore, it is imperative that a strategy be developed as to how the development of nanotechnology can be organized to both improve the lives of people in industrial states as well as advance living standards of those in developing countries. Also within individual countries and societies a sharp nano-divide between the well off and the underprivileged ought to be prevented. It would clearly be problematic from the viewpoint of equity, if further development of nanotechnology would dispro-portionately benefit the higher echelon of society. In that case existing gap between haves and have-nots would only widen.

Privacy

The extreme miniaturization that nanotechnology facilitates would be enormously advantageous for further developing espionage and surveillance gadgets. It is not hard to imagine infinitesimally small tracking devices, nanocameras and nanomicrophones registering virtually everything we do without us even noticing it. Evidently, this idea poses the question of whether, and if so how, we will be able to defend our privacy.

Playing God

From a religious point of view, it could be argued that fully advanced molecular manufacturing would demonstrate a problematic attitude towards God's creation. After all, it would involve a fairly fundamental reshaping of creation in order to comply with human ideas and directives. Taking God's creation apart in the most extreme form and putting it together in a molecule-by-molecule way is likely to offend those who believe that the structure of creation is already pervaded by divine rationality and heavenly benevolence.

Approach to nature

In a way that would not necessarily have to be based on religious premises, a similar argument could be constructed around the attitude to nature that is involved in molecular manufacturing. Constructing our surroundings in a molecule-by-molecule way demonstrates an invasive and instrumental approach to nature that could be seen as conflicting with the idea of respect for nature. The idea that nature has finality and some kind of moral meaning is a very old tradition in western philosophy. It can be found already in the philosophy of Plato, Aristotle, the Stoics, the Neoplatonists and Thomas Aquinas. More recently Goethe, in his theory of colours, criticized Newton for having forced light through prisms, thereby demonstrating a wrong approach to nature and thus coming up with biased results as to the real essence of light.

Surmountable Ethical Problems

After having determined the ethical problems that are connected with further developments in a specified nanotechnological field, analysis should decide whether they can be dealt with or not. After all, if it happens that the ethical problems are not surmountable, it seems ethically objectionable to favour further developments of the research.

To exemplify this last step, an example is given that concerns risk management. To reduce the risks of the further development of techniques of molecular manufacturing using assemblers the Foresight Institute has issued the so-called "*Foresight Guidelines on Molecular Nanotechnology*". These guidelines contain a set of "*Development Principles*" as well as "Specific Design Guidelines". Among the first are prohibitions such as the statement that artificial replicators must not be capable of replication in natural environments or directives reminding developers that they should attempt to consider systematically the environmental consequences of the technology and limit them to intended effects. Among the "*Specific Design Guidelines*" are directives focussing on ways to device designs to prevent self-replicating devices running amok such as complete dependence on artificial "*vitamins*" not available in any natural environment or programming termination dates into devices.

Hardly ever has there been such a discrepancy between opposing evaluative judgments as can be observed in the debate on nanotechnology. However, both the radically optimistic as well as the thoroughly pessimistic ideas seem to be founded on a somewhat one-sided and narrow approach in ethical assessment. Nanotechnology is a complex endeavor that consists of many different projects with a huge variety of goals. Hence, different fields of nanotechnological research should be distinguished.

After all, these fields will not necessarily demonstrate identical or even similar ethical aspects. Up to now, most of the debate has been too emotional, too general and quite decontextualized. In order to improve ethical assessments of nanotechnology, a six-step method has been presented. Use of this method results in a rational and systematic assessment of the ethical desirability of further development of a specific field of nanoresearch. Due to the complexity of nanotechnology, it will be necessary to create interdisciplinary groups for ethics research. These, in turn, would need adequate funding. Therefore, it would be a good idea to devote a certain percentage of the government spending for nanotechnology to ethical research such as we have done in earlier attempts to address the ethical, legal and social implications of the human genome research.

Politics of Technology

The Foresight Institute, with which the public is most familiar, is oriented toward educational and informational activities. Its broad task is to make people aware of the concept and possibility of nanotechnology, and to promote discussion on the topic. The Institute for Molecular Manufacturing is the *promoter* and *sponsor* of technical research in molecular nanotechnology. Its function is to increase the likelihood that technical work is performed in this country, and that such research is directed into useful approaches. The Center for Constitutional Issues in Technology (CCIT) is the public-policy-oriented organization of the Foresight Family, and its purpose is specifically to promote discussion on public-policy issues, and (*analogously* to the aim of IMM in the technical-research sphere) to attempt to influence thinking in the public-policy sphere. CCIT wishes to ensure that discussions on the public-policy issues engendered by nanotechnology occur, and that they go in directions that we think are productive.

CCIT was founded in 1991. It has spent the intervening time in organizational tasks, in defining its areas of research, and in preparing potential research programs. Currently CCIT has prepared an initiative regarding a near-term issue—standards in emerging nanotechnology. This program would examine some of the questions regarding product standards in emerging nanotechnology and molecular manufacturing, comparing governmental versus industry roles in emerging standards.

CCIT would also like to examine some long-term issues regarding equitable access to resources, an issue that, as you can tell from the discussions presented in this book, is an idea that has drawn considerable interest in the context of the emergence of nanotechnology. Recently, CCIT also has begun to define an initiative to look at international security and examine potential concerns regarding the military security issues in nanotechnology. This initiative would examine some proposed control regime models, considering both the possible benefits and the possible hazards of such control regimes.

CCIT feels initiatives will likely create international controls on the proliferation and use of nanotechnology in consequence of the weapons implications of the technology. CCIT is quite concerned that, in the course of creating such controls, the benefits of nanotechnology could be severely circumscribed or denied entirely. CCIT is also concerned that there could be serious infringements on the ability of persons and organizations to experiment in seeking new directions for society, or conversely in attempting to conserve existing values.

CCIT is also concerned that certain trends and developments might give rise to international conflicts. The approach that we favour in meeting the potential problems of a nanotechnology era is one in which a protocol for interaction between people is established. This protocol would maximize individual rights and liberties, while also attempting to establish the minimum effective types of control to prevent major harmful consequences to the world, which could (at their worst) be very serious indeed. However, in seeking to determine the nature of protocols that will serve these ends, the only ones that people will accept in the long term are fundamentally constitutional ones—that is, ones that balance whatever protection may be needed against people's liberties, in a predictable fashion with appropriate safeguards against abuse over time.

It is not possible to discuss the *politics of nanotechnology* today. As of today, there are no politics of nanotechnology. There is long-range speculation regarding a possible politics of nanotechnology. However, given the way politics works, there are not going to be any issues in nanotechnology until we get closer to an actual *nanotechnological assembler* or *proto-assembler*, or some kind of actual device.

The governmental means and ends and mechanisms that exist today to deal with technology will likely be the ones that will deal with nanotechnology when it emerges. Defense analysts have a term, "*come as you are party*," that refers to the fact that when war starts, one fights with the weapons that are in place at that time. The lead time to build and deploy a modern weapons system is considerable. Similarly, nanotechnology policy will be a "*come as you are party*," in that the *institutions*, *laws*, and agencies that will address nanotechnology policy and regulation, will be the ones that are in place before anyone takes it seriously—that is, the ones that are in place today or get created very soon.

A fundamental division between the forecasting function and the active functions is the funding of research, and the policy-and-regulatory-formation function. The former consists of the parts of the United States government chartered to perform long-range forecasting and to examine new technologies on the horizon, to consider what policy questions they might engender, how they might be addressed, and to make recommendations regarding appropriate courses of action.

Unfortunately or fortunately, depending on your point of view, these views and recommendations usually have only very limited impact on the actual emergence of the technology, or the actual policy measures adopted when those technologies emerge. In actuality, no matter what the long-range planning recommends, when the issue actually comes before Congress and the Administration (in this discussion, when a company or an institution creates an actual nanotechnology-related device), that is the point at which one must consider the issue to be raised *in vacuo*. None of the long-range forecasting previously performed will necessarily constrain the Administration or Congress regarding the assignment of regulatory responsibility or the nature of the applicable regulatory regime. Other, more short-term considerations addressed in this chapter have a much greater effect on such issues.

A Lesson from Space Policy

Some prior existing actions can bind, or at least circumscribe, the outcome of the policy and regulatory process. The commercial *space field* offers instructive parallels. Certain international treaties negotiated in the late 1950s and early 1960s (and finally signed and ratified in 1967), along with some implementing protocols added in 1972, did set a number of bounds on subsequent commercial space activity. Specifically, the Outer Space Treaty of 1967—a multilateral United Nations treaty between the United States, the former Soviet Union, and the United Kingdom, France, and other countries—contained a clause that effectively held that private activity in space was legal. In fact, President Kennedy delayed the original signing of the treaty in 1963 until the first commercially owned and operated satellite was launched, in order to ensure that a precedent for private property in outer space would be established before the treaty was signed. At that time, the former Soviet Union's position was that any private property in space should be forbidden in the treaty. The final text of the treaty did not accomplish this. However, as a compromise, a clause was inserted in the treaty, to assign unlimited liability for damage caused by objects in space to the "*sponsoring state*," a term that was defined imprecisely and permitted at least two differing interpretations. Subsequent legislation continued the ambiguity. As a result, when first began efforts to launch rockets on a private basis as entrepreneurs, the government's reaction was to say, in effect, "The Government has a direct stake in your activities. No matter what you do, the government has the liability for it, so we have to take a heavy role in your insurance requirements and all that drives them."

As a result of these prior treaty constraints, it was not possible to follow a maritime or aviation precedent in third-party liability issues in space commerce, although such a precedent might have permitted a more favourable outcome. The lesson for other emerging technologies is that it is possible to have international agreements negotiated well in advance of the advent of the technology. Because these agreements cannot accurately foresee the circumstances of the emergence of that technology, they can result in a binding of the hands of those attempting to create the regime at the time of emergence. In other words, bad foresight may be worse than no foresight. Currently existing international agreements (particularly regarding environmental issues, or in munitions control and *weapons proliferation*) may affect nanotechnology, and it is not even clear at this time which agreements those might be. When the technology emerges and we can actually ascertain its nature, and better know what the first incarnation of this technology will be, then we may discover that some of the basic laws of nanotechnology have already been written and agreed to. That will be more clear as we get closer to that time.

Players in the Government

The study or forecasting function is its own universe. Similarly, the active side divides into a number of universes. Among these, it is particularly useful to examine two sets of players: one are the sponsors of research, and the other are the policy and regulatory players. The research universe will be one of the first areas that impacts nanotechnology. The first issue within the research realm is allocations for basic and developmental research. (Given that Japan has begun significant funding for such activities, the question of why the United States government does not have such funding becomes an active, rather than a speculative question.) In the policy/regulatory realm, it will be the regulatory activities that, as commercial nanotechnology begins to evolve, will begin to affect those efforts.

In the research realm, the mainstream science research establishments (such as the *National Science Foundation* and the *National*

Institutes for Health) allocate most of the civilian research funding. Neil Jacobstein's presentation on the federal research establishment constituted an excellent and detailed survey of this arena. Much basic research is done under the aegis of the Department of Defense. However, much of this research is not *applied research* (research that is actually developing or building weapons). Much is *basic research* because they realize that unless there is an understanding of the basic physical principles underlying technology, advanced weapons development must be limited. The Defense Advanced Research Projects Agency (DARPA), which is now the Advanced Research Projects Agency (ARPA), sponsors a wide range of research. Some of the service laboratories, particularly the Naval Research Laboratory, sponsor much basic research. There is also substantial intermediate development work performed at such laboratories, which is as applicable to commercial technologies as it is to military technologies. Laboratories sponsoring such intermediate research could be early participants in nanotechnology research. Some important precursor research relevant to nanotechnology is currently being sponsored at the Naval Research Laboratory.

A beneficial aspect of the defense research arms is that a substantial amount of the basic and early stage applied research is done in a nonclassified environment. Over the last decade, the Strategic Defense Initiative Organization, or SDIO (the so called "*Star Wars*" organization, which is now the Ballistic Missile Defense Organization (BMDO)) has sponsored interesting basic and applied research, much of which has been good research. An interesting aspect has been that SDIO has been more open to different and unusual proposals for research than many of the established research institutions, as they relied more on the individual discretion of the granting officers. This has been especially true in their "Innovative Science and Technology (IS&T) Program," which has been a source of support for many innovative high-technology entrepreneurs over the last decade. Much of this research has very little to do with the ultimate "*Star Wars*" system, but rather are interesting pieces of technology that some officer decided would be a good thing to do.

The White House offices impact both basic research and long-range planning and thinking. The Office of Science and Technology Policy (OSTP) three decades ago was the President's science advisor. Today, the Director of OSTP is called that, but originally he literally was that (that is, a scientist whom the President would occasionally call to ask for advice). In the way of Washington, they presumed, "Well, you have to have a secretary." So, they got him a secretary, and then, "Well, he ought to have an assistant." Before you knew it, there were 30 people in an office. It was enshrined in the Old Executive Office Building (next to the White House), and then they had to spill over to the New Executive Office Building. Soon it was bigger than cabinet departments were in the day of Thomas Jefferson. The bulk of what they do is forecasting and thinking, and they issue papers, and it does affect decisions.

The Office of Management and Budget (OMB) has, in reality, more impact because they have the final say on the Administration's decision regarding what gets funded. The Program Area Director for Science, Technology and Space, inside the OMB, has much more to say about what the Administration will and will not do in space and science than OSTP. People often overlook that fact, but it is well-known. It is the "*Golden Rule*": the one who controls the gold makes the rules. This situation is no exception.

In considering the Congressional side, one must remember that under the American system of government, the President proposes, but the Congress disposes. Congress makes the final disposition of funds. There are programs that agencies never wanted, but that Congress tells them they are going to have any way. The Osprey V-22 vertical take-off tilt-wing aircraft, which the Pentagon never wanted, is an example. The decisions in Congress are, in essence, made by a few key Congressmen and their key staff members, especially the permanent committee staff that report to the chairman, as opposed to the staff of an individual congressman.

The authorization committees will say, "We need a program in X, Y & Z. Let us allocate $100 million to this program this year, and we will allocate the money to be spent in District A and District B." To accomplish that, the authorization committee generates an authorization bill that authorizes that money. However, that money is not real until the appropriations committee then says, "Yes, we will spend the money," or, "No, we will not spend the money." Their effect, on the Congressional side, is somewhat equivalent to that of the OMB in the Executive Branch process. Of course, the full House and Senate must approve the decisions of both authorization and appropriations committees (just as the President must approve the decision of the OMB), but a failure on either part to make such approvals is a rare and remarkable event.

The Office of Technology Assessment (OTA) is an independent office reporting to Congress (as opposed to the Executive Branch), which performs long- and short-range studies. Their charter is to examine any particular technology question and draw up scenarios that attempt to foresee consequences of alternative courses of action. In effect, they generate stories that say, "Well, if you do X, this will happen. If you do Y, that will happen." They are not sup posed to make recommendations, although often you have a report that, in essence, says, "If you do X, it will bankrupt the nation. If you do Y, you will blow up the world. If you do Z, everything will work out great, but we are not making any recommendations. Go do what you want."

Analyzed operationally, one might be tempted to say that the primary function of the OTA is to provide jobs for technology-oriented staffers of the party not holding the Presidency, since they cannot get any jobs in the Executive Branch when their party is out of power. Following this, one might predict that you will now see a migration of Democrats from the staff of OTA over to actual operating agencies.

All of these elements affect the outcome in varying degrees. The locus of early research may determine some elements of the ensuing governmental reaction. For example, if a substantial portion of basic nanotechnology research is funded by NIH, that research will be oriented toward medical and health applications. If it is funded through NSF, it will be more generic. If it is funded through the National Institute of Standards and Technology (which is part of the Department of Commerce), then it will be probably more commercial- and manufacturing-oriented. If it is funded through DARPA, it may very well be oriented toward defense applications. It is quite possible that there will be parallel programs running in all those places, as well as independent research financed in the commercial sector.

Looking at the executive departments themselves—those that are headed by cabinet Secretaries or Administrators (as is NASA, which is an independent agency)—a number of the different cabinet departments have substantial interests in technology. The Department of Commerce has the National Institute for Standards and Technology, formerly the National Bureau of Standards; the Patent Office, which has its own independent and very strong impact on the way technology develops; the Bureau of Export, which determines what can be exported and what can not be legally exported (aside from the weapons-export issues addressed by the State Department rather than by Commerce); and the Department of Energy which runs National Laboratories such as Livermore and Los Alamos.

The *Department of Energy* has the Federal Energy Regulatory Commission (FERC), which regulates many energy-related activities. If nanotechnology first emerges as part of an application in an energy production area, FERC may be the agency that does the first regulations on it.

The *Office of Munitions Control* (OMC), part of the Department of State, administers the International Traffic in Arms Regulations, which control many technology enterprises not normally thought of as weapons (such as commercial communications satellite technology). OMC is one of the leading candidates for being the controlling agency for nanotechnology. It must be understood that "export" in this sense is not limited to taking a device, putting it in a crate, and shipping it overseas. It can mean putting the plans or information relating to a device or technology in an envelope and sending it overseas. Or, it can mean taking information that is critical to developing a technology and publishing it in the United States, since once it is published here, it is available around the world. OMC, by this logic, has an enormous amount of power. Some speculation has been published in the legal press that the OMC's charter and the Munitions Control Act of 1950 may, in fact, be unconstitutional in that they violate the First Amendment. That point has never been put to a firm test. It may be soon.

NASA is an interesting anomaly because work that would ordinarily be performed under the aegis of NSF (if it is space-related) will often be performed by NASA.

The *Environmental Protection Agency* (EPA) has an enormous amount of regulatory power and, since the release of nano mechanisms in the environment also potentially falls under the jurisdiction of the National Environmental Policy Act, EPA could end up being the regulatory agency.

Since the *Federal Drug Administration* (FDA) has jurisdiction over medicine and medical devices, if the first applications of nanotechnology are in medicine or medical devices, they may be the lead regulators. As John Doerr has said in his presentation, it takes about a $100 million to get a drug approval through FDA. If a nanotechnology device is interpreted, as it may be, to be a drug rather than a machine, then this could mean that the product will not be seen first in the United States, but rather elsewhere.

To emphasize the general implications of the previous point, when the first *nano device*, *nanomachine*, or *proto-assembler* device comes on the market, the regime that is going to control it is probably already in existence. It is not clear which one it is going to be, but the three top candidates at this time would be the EPA, OMC, or the FDA. None of those is the most suitable, in my opinion, but quite often, from the standpoint of businessmen, if the regime is already in existence and already understood, that is a plus. You can go to a lawyer and get an estimate of how much it is going to cost you, and how long it is going to take to get an approval through. It may be that if a company has a nanotechnology device that is close to market and has (let us say for the purposes of discussion) a medical application and a manufacturing application, the company may (out of caution) make a decision to bring out the manufacturing application first because the regulatory path is easier. Such considerations will be real, and they will arise.

Now that Al Gore, who held hearings on nanotechnology in the summer of 1992, is Vice President and head of several important councils, it is possible that nanotechnology will begin to emerge as an item of discussion. One of the actions that has been discussed was that of having the OTA include nanotechnology in its study areas. With Vice President Gore's party in power, this task may shift from OTA over to one of the Executive Branch agencies.

The government will not do much beyond forecasting and perhaps sponsoring of long-range research until an actual company with an actual project (or alternately, an actual federal agency with an actual internal project) is about to do something. The agency with the most proximate charter will have the strongest incentive to act. There may be a little "*gold rush*" of agencies at the time when nanotechnology approaches implementation, or maybe a big frenetic rush, depending on how important it is seen as being.

Certainly, when commercial space emerged as an issue in 1981 and 1982, there were prospectors from the Department of Commerce, the Department of Transportation, NASA, FAA, Bureau of Alcohol, Tobacco and Firearms (in its capacity as the regulator of explosives), and about 23 other agencies (including Department of Agriculture and Department of Interior), all investigating the possibility of asserting jurisdiction and preparing arguments to support their cases.

Influencing the Government

In influencing the governmental reaction to *nanotechnology*, CCIT would like to do a number of things. When the "*gold rush*" starts, we can have a considerable impact by evaluating the various options for regulation (or approaches to regulation that are presented by the situation at hand) and trying to work with the agencies we think are most appropriate. This is a classic tactic in such situations. It may be possible to propose that a new agency be established within one of the existing cabinet Departments and to promote the idea of that agency being created and given the jurisdiction.

That course of action has certain benefits, because in doing so, it creates a little more leeway in suggesting what the agency's regime is going to be like. A party active in the field has a lot of leeway in suggesting what the regulating regime is like, no matter who gains jurisdiction, because typically you have some bureaucrats who have just been given a new role. They want to make the industry happy. They do not need the industry going up before Congressional committees every six months and saying that this decision is a disaster and that the agency in question ought to have their jurisdiction taken away. The agency does not desire this. They need the industry to say, "These people are doing a great job and their budget should be increased." So, they will listen to industry. It is a "*mutual capture*" process.

The agency will let the actors in the field write the regulations to some extent. If you pay to have the legal work done, to have a draft regulation written and present it for the agency's edification, the ideas and approaches presented will often get substantially incorporated in the draft regulations that are created. Everyone likes to have someone else do work for them, and bureaucrats are no exception.

This is not necessarily a *pernicious process*. In reality, many people in the federal government realize that people out in industry (or out in the front lines of the research) know much more about their work than they themselves do, and they listen to suggestions. From the public interest perspective, the problem with this process of mutual capture is that it cannot prevent the cooperation from including cooperation at the expense of potential new entrants to the field, and thus at the expense of the public.

CCIT is planning to pursue multiple, complementary courses of action. For the present, CCIT will be active in the forecasting world, CCIT is seeking to influence this realm by advancing its scenarios. Because of CCIT's status as member of the Foresight family, and, thereby, its access to the best, most cutting-edge knowledge on nanotechnology, CCIT is able to have the most accurate understanding of the prospects for this area, and, therefore, the most useful scenarios. As the research advances, CCIT will foster discussion of what kinds of research are done, what kinds of institutions are funded, and by what means funding is allocated. Subsequently, when actual regulatory questions arise, CCIT intends to be present with suggestions on what to do.

Industrial Policy

To give an example of what technology policy debate is like, we want to make a few comments about a topic which four or five of the other presenters here have mentioned a lot, but about which there has not been much substantive discussion. That lack is not the fault of the presenters, because this issue has been discussed mostly in a circular fashion for the past fifteen or twenty years in Washington. This is the so-called "*industrial policy*" issue.

To begin with, it is quite a bit of a task to define "*industrial policy*" precisely. At its simplest level, industrial policy is the set of governmental policies that affect industry, including the kinds of issues that have just discussed—funding levels, mechanisms, and categories for basic and applied research, regulatory regimes, and similar issues, such as been advocating that the government do, or in some cases, not do. However, more recently, another, more specific understanding has come to the fore in the debate—industrial policy in the sense of an integrated government plan to foster desirable industries and technologies.

Many people have advocated industrial policy in that sense recently, especially since the Clinton Administration (which has promised to have an industrial policy of that nature) has taken office. There is an attitude that says, in affect, "Now that we have an industrial policy, we have made it. Our problems are over." In fact, once it is determined that the nation will have an industrial policy, in many respects your troubles have just begun. Because you must define exactly what kind of industrial policy you have, you must define how you are going to run it and which course among many you will take. It is important to keep in mind that the success or failure of many, if not most, technology companies or initiatives will now hang on the nature of these decisions.

Advocates of industrial policy often point to foreign successes that they attribute to successful industrial policy. However, industrial policy has had a very mixed track record overseas. How many of you use a computer of any sort in your work? Please raise your hands. (everyone in the audience raises hands.) How many of you use a Machines Bull mainframe as your computer? (hands are raised.) Nobody? How many of you even know what "*Machines Bull*" is? (hand is raised.) Machines

Bull is the European "*national champion*" computer company. This was the product of pan-European industrial policy, and particularly French industrial policy. The European governments, and particularly the French government, poured billions of dollars of French and European resources into trying to make Machines Bull an equivalent competitor to IBM. For various reasons, this effort was, in essence, a complete failure. The French government still uses many of these machines, and, to some extent, European businesses do, partly because they get pressured by the government to do so. One must ask, "What is the opportunity cost of putting all that money into a massive failed project? What could have happened if that money had been left to industries to spend for themselves, or spent by the government in a different fashion?" Industrial policy is not an automatic success. It has had quite a mixed track record overseas.

One must note, and wonder about the Clinton Administration's promise of both strong industrial policy and strong actions on creations of jobs. They have had a heavy emphasis on job creation. Countries with strong industrial policies often are very poor in job creation. The continental European states are considered one of the classic examples of strong industrial policy. They have higher unemployment rates (particularly for young entrants in the job market) and they have had a much poorer record on job creation over the last 15 years than the United States. This includes France, where they artificially maintain a high percentage of people in the agricultural sector relative to other industrial nations, thus absorbing a lot of people that would otherwise be in the industrial job market.

Japan has a structure that absorbs a significant number of people, not through deliberate industrial policy, but through custom (which maintains an inefficient retail apparatus), as well as a highly subsidized agricultural sector. The result is that many Japanese are underemployed at low wages. Those people are being held off the industrial job market, plus, because of Japanese social customs, most women are kept out of the job markets. It is not clear if they had not maintained those social differences that they would be much better than the United States in terms of unemployment. It is not clear that in cases where countries have industrial policies and strong successful competitive industries, whether the successes are because of the industrial policy or other factors. It is the old question where rain and wet streets are associated, but you cannot make it rain by wetting the streets. So, you must be careful in attributing cause and effect in examining industrial policy.

South Korea has a strong industrial policy with central planning. They have had highly successful industries in automobile, shipbuilding, and a number of other areas. Taiwan and Singapore have had very different types of industrial policies, but they have also been very successful in job creation, competitiveness, and good foreign trade balance. One might argue that these are examples of success through strong industry-government cooperation. These countries have gone from being at the income level of Haiti in 1960, to having giant foreign trade reserves and now pressing into the ranks of developed nations. So, clearly there is a correlation—or is there? Consider that South Korea, Taiwan, and Singapore are three of the so-called "four tigers." The fourth "*tiger*" is Hong Kong, which has had an equivalent (or greater) rise of success in competitiveness, but has had no industrial policy whatsoever. So, it is not entirely clear that the successes are because of the industrial policy practiced by those governments.

Another factor to consider is the legal and cultural systems of those countries. It is worth noting that the countries with legal and cultural systems most similar to the United States that have tried to implement industrial polices have had the least successful outcomes from it.

Canada probably has the political, legal, and cultural system closest to the United States, and it has had a very dismal record in industrial policy. The United Kingdom has had a strong industrial policy for 30 or 40 years, and, in addition, has poured a considerable amount of money into science and technology education. They still have excellent universities, excellent education programs, and they have been totally unable to turn this advantage into competitive technology or products. One must ask, "If we are to have an industrial policy in the United States, what aspect of our law or culture might we have to change to make this more like countries in which it has seemingly been more successful?"

Our liability *law system* creates many hazards and traps for companies putting forward new technology. It is not clear that we could have a successful industrial policy unless our liability policies were more like that of Japan, Germany, and France (that is, tort reform). However, will we have reform in that direction? Consider that trial lawyers, who have formed a well-funded lobby against tort reform, were the largest early donors to the Clinton campaign. This fact suggests that tort reform of the sort needed to support industrial policy, is unlikely for the current presidential term.

Antitrust is another area. We have an *antitrust system* that is unlike that of any other country in the world with strong industrial policy, and it has in the past worked against the kind of industry-to-industry cooperation (much less industry-to-government cooperation) that is typical of industrial policy systems. Are we going to change the antitrust system? This is not evident. Historically, the Democrats have been the pro-antitrust legislation party.

Continuity in government is another area. Typically when Europeans are promoting an industrial trade product, their trade officers and their project officers remain the same despite changes in party or administration. Britain or France typically change about 300 civil servants at the time of the transition from one party to another, and their parties are more extremely polarized than those in the United States. The United States Government typically changes about 3,000 officials. We have had space trade negotiations with the European Community, China, and the former Soviet Union, which are all extremely important to the success of the United States launch industry. Do you know how many people in the United States government have been with those negotiations since the beginning, and have a full, continuous understanding of the issues? One—exactly one man, and unfortunately for the prospects of continuity in this area, he is a Republican, so he is unlikely to retain his position. He is very likely to be replaced. If we have an Administration that advocates continuity in industrial and trade policy, should they not be keeping him? Should they not be keeping the person with the long track record?

Another example can be seen in the negotiations over the *Airbus airliner*, where the United States has been negotiating with the European Community since the late 1970s. The key person at the Aerospace Industries Association who had the responsibility for tracking those negotiations. She observed, "The Europeans have had the same team from the beginning up until today. The United States has switched teams every time they switch an Administration, and quite a bit in between." So, are we going to develop the tradition of continuity in government necessary to make this continuity possible? It remains to be seen.

"*Research monocropping*" is a problem. Just as a *monoculture* in forests or other agricultural areas is bad for the ecology in the long-term, which can be defined as "a tendency to concentrate all your research dollars on a common set of programs, a common set of directions." This is very bad for research. Monocropping in research means that you do not try the different veins of ore. You do not try the different paths. However, the incentives in government funding are toward research monocropping, because there is a prevailing assumption that the government plan is going to be able to identify all the productive areas of research, and fund them all. The ones that do not get included in the master plan, by definition, do not get the resources.

Right now we have considerable diversity because the research activities in different parts of government are quite uncoordinated. The beneficial effect of this is that it permits different directions in research. NSF allocates its research funding by peer review. DARPA largely leaves it up to the discretion of the program managers in the areas, so it goes by individual judgment. SDIO IS&T basically does the same thing—they give even more leeway. If there were a "*Rationalized Plan*," there would be a danger that the different approaches funded for very different reasons might all be brought into a single coordinated plan, which would mean there would be less room for different ideas.

Another problem is leader-versus-follower strategies in industrial policy. Industrial policy has historically been better for followers than for leaders in technology. The South Korean government said, "We want a steel industry." It is relatively easy to determine what goes into a steel industry. One needs steel plants, limestone, this and that, but it is all well-known. One can copy an existing foreign industry, incorporate the best of known current technology into it, and know how to put that together. One cannot say, "We want a nanotechnology industry," and proceed to put one together, because one cannot know yet what is going to be necessary for a nanotechnology industry. There is no model to copy. The states that have been most successful in industrial policy have generally been states that were catching up, and that were going into an existing industry and following the lead of other nations. There have been very few examples of nations that have gone out and seized leadership in a pioneering technology industry through industrial policy.

It is worth turning to Japan briefly because many people have an idea of Japan as a country with a successful industrial policy, which is a bit reminiscent of the caricatures of Japan that occurred during *World War II*. Because Japan enjoyed all these tremendous military successes for month after month during the first two years of the war, many people assumed that there was a Fu Manchu-type central genius in Japan who was planning the whole war, through which everything was tightly coordinated, and which provided that the Navy would strike here and the Army would strike there. After the war, when we were able to read the actual plans and interview the officers, and after participants published their memoirs, it became apparent that the Army and the Navy were virtually independent of each other and of any kind of central government coordination. Both were proceeding adventitiously. Admiral Yamamoto, perhaps the most perceptive officer on the Japanese side, made a famous statement at the beginning of the war. He said, "I can run amok for two years. After that, you had better negotiate." In reality, there was no plan. Japanese forces probed adventitiously in every direction. When they saw weakness in a target, they pushed, and it fell.

To some extent, Japanese industrial policy successes are like that. *The Enigma of Japanese Power*, as a general discussion of Japanese social and political institutions relevant to the industrial policy debate. A number of different, strong institutions exist in Japan, and strong coordinated sectors of government exist in industry, but they are not universal. There is no Japanese master plan. Japanese companies, and secondarily governmental institutions, are pouring money into a diversity of efforts and directions. Some of these have been very successful, and some of them have been great failures. To imagine that MITI is this same kind of omniscient central planner who knows all, sees all, and plans all, and is responsible for all these victories, is just as fallacious as it was during World War II. The world industrial market, at the level addressed by Japanese companies, is too complex to be dealt with by a central plan.

What many industrial policy advocates have been presenting is a kind of cartoon caricature of Japanese industrial policy, which is what some advocates would like to see implemented in the United States. It does not necessarily have that much correspondence to what is going on in a country like Japan. In a lot of ways, South Korea has a program that is like what people imagine Japan's to be, but South Korea deals much more with catching up on old technologies than with new technologies, and has historically chosen to compete in many fewer sectors than Japan.

In closing, it is also worth raising the issue of the Iron Triangle problem—the problem of political allocation of funds as affected by the interlocking interests of local beneficiaries of government funding, their representatives in legislative institutions, and the Executive Branch persons who depend on that legislature for funding. When new technologies are emerging in many new areas, one must invest in multiple paths. One must be prepared to switch support from area to area rapidly, because one can be pursuing a research path, and continuing on that path, and it will appear very promising. One will be mining the vein, to use Howard Landman's metaphor, and it appears to be fruitful, but then suddenly it dwindles, while an expected little side vein actually proves to be the route to the mother lode. In such cases, one must be able to switch one's resources from the dry hole to the promising vein.

When you have a central plan, this is hampered, as researchers become invested in a particular approach, and people began to obtain benefits, in effect, by mining the dry hole. The money will go to the research institute that is taking a certain approach and happens to be located in the Congressional district of the chairman of the committee. If that research area is becoming unpromising, one may need to reprogram resources to a different institution and a place that has the experts in a very different, desired area. Consider, then, the situation of the master planner who must say, "We must adjust the master plan. We must reduce the funding in Mr. Chairman's district, and we must reprogram it over to the district of Mr. Nobody." To say the least, very strong incentives exist against that happening.

The United States, government could not spend any money productively on technology. The government can do intelligent things in the way it spends money, especially by being a smart customer in buying those things the government actually needs for its own purposes, which can do a lot to stimulate useful technology and technology development in industries in the right areas. However, we should think very carefully about spending such money. My purpose has been to raise some points and, in doing so, to make clear that industrial policy is not a magic wand, or a magic solution to all problems. Before one could expect to create a useful industrial policy, many real problems must be discussed. CCIT intends, among its other tasks, to promote discussion on such issues.

5

Revolutionary Nanotechnology

Roughly seven thousand years ago, humans began to leave their nomadic ways and form civilizations around the irrigation and cultivation of land. As a result, human society and community transformed radically. The creation of government and bureaucracy, of social classes, written language, the rule of law, the notion of the individual, standing armies, and much more, all emanated from this technological change. Dubbed the "*irrigation society*" by renowned management thinker Peter Drucker, this first great technological revolution of man lasted over two thousand years. Nowadays, the word revolution is used rather freely. From the "*internet revolution*" to the "*digital music revolution*" to the "*nano-technology revolution*," at what scale does an innovation become more than an innovation? In her classic work *Technological Revolutions and Financial Capital* Carlota Perez defines five technological revolutions since the end of the eighteenth century.

As new technologies emerge and disseminate, they tend to follow similar economic investment cycles which Perez calls "*techno-economic paradigms*." However, the key realization – and what Drucker was suggesting in his 1965 presidential speech to the Society of the History of Technology – is that we are living in a second great technological revolution. Beginning with the Industrial Revolution in Britain (around 1750) a man could expect to die in a world very much different from the one into which he was born. A great revolution implies an increasing rate of innovation, and not just technological innovation, but also organizational and political, in many different fields. There is little doubt that the rate of innovation accelerated after 1750 and is still increasing. Although it is difficult to assess the evolution and longevity of such a revolution from within it, we can easily believe that as more people and more wealth come into the enterprise of innovation, more of Perez's techno-economic paradigms will occur with increasing frequency. Nanotechnology may be one of them.

Technological revolution is always accompanied by political and social change. The two go together. The bigger the technological change, the more society must adapt to accommodate this objective reality. For example, as farmers began to accumulate wealth (that is, food) for communities, an army became necessary to protect it. Just as there are larger and smaller technological revolutions, there are larger and smaller societal changes to go along with them. In Table 5.1 we have tried to form a hierarchy of revolutions. Inevitably, revolutions are coarse-grained, representing a sum of individual innovations that erupt seemingly randomly. We define four categories of revolution: great revolutions, of which there have only been two; with the present one just beginning; major revolutions or the techno-economic paradigms of Perez, which usually last some 50 years; minor revolutions, which are finer still and are key building blocks of the major revolutions; and micro revolutions, which are largely new investment opportunities in technology that come about within the larger revolutions but also follow a cyclical pattern of investment and saturation.

Soon after World War II, governments worldwide and particularly in the US realized that significant investment in the natural sciences could drastically affect the power and wealth of nations. Integrated electronics, the Internet, even lasers can be lumped into the age of information and telecommunications as a direct result of this investment. Much of what is today called nanotechnology naturally follows from these lines of technical pursuit and scientific inquiry. Together with the theoretical understanding of quantum physics and electrodynamics that has developed over the last century, this continuation of research has led us within reach of tremendous rewards from the manipulation and control of matter at the nanoscale. Whereas these pursuits were somewhat ignored in recent decades – significant hype and investment focusing instead on telecommunications, software and networking, and biotechnology – there is a growing realization that it is time to start pushing materials science and fundamental research again. Very recent trends in the global energy crisis and the so-called green revolution only add to this notion.

Table 5.1. Hierarchy of technological revolutions

Great Revolutions (there have only been two that we know about)
1st: Irrigation society, began–ended: approximately 5000 BC-3000 BC
2nd: Began with the Industrial Revolution in Britain in the 18th century
Major Revolutions after 1750 (start date)
the industrial revolution (1771)
the age of steam and railways (1829)
the age of steel, electricity and heavy engineering (1875)
the age of oil, the automobile and mass production (1908)
the age of information and telecommunications (1971)
the age of bio-engineering (1980)?
the second industrial revolution (1991)?
the age of machine-phase nanotechnology (2030-50)?
Minor Revolutions (some examples)
personal computing
mobile phones
global networking
nanoparticle revolution?
Micro Revolutions (some examples)
digital music revolution
HD TV revolution
nanoparticle revolution?

If anything, nanotechnology has become a marketing term to encompass and drive this belief that more funding is needed in the physical sciences to maintain economic, scientific, and military advantage over international competition. As evidence, roughly one-third of the budget for the *National Nanotechnology Initiative* (NNI) this year will go to the *National Science Foundation* (NSF) (Roco, 2004), which primarily supports unfettered basic research. Still, understanding how government and the military drive technological development and how nanotechnology as it stands today (and may exist in the future) may relate to prior revolutions in history has great value. Responsible encouragement of the great technological revolution in which we find ourselves is vital to human civilization.

Defining Nanotechnology

Nanotechnology is a social construction. The word nanotechnology did not emerge as a distinct area of science, but rather was introduced externally and defined by its usage in the greater societal dialogue. A primary consequence of this very public defining of the term nanotechnology is it's present bipolar nature. We take a pragmatic definition of nanotechnology that combines both sides: the *reality* of the word as it is used today primarily by governments, corporations, and scientists as well as the *vision* of what the field might become. The *reality of nanotechnology* – is defined mostly by government funding managers and agencies – largely encompasses ongoing research in materials science and solid-state physics. Examples include *nanoparticles* and *quantum dots*, "nanoenabled" surface coatings, transistor features that are less than 10 nm scale, giant- and colossal-magneto resistance (as in hard drives), spintronics, photonic band-gap structures, and more.

The definition usually takes a variant like this one from the Royal Society of the UK: "*nanoscience* is the study of phenomena and manipulation of materials at atomic, molecular, and macromolecular scales, where properties differ significantly from those at a larger scale." Distinct from this is the more science fiction *vision* of nanotechnology popularized by Eric Drexler and in books like Neal Stephenson's *The Diamond Age*, that of atom-by-atom construction of matter and nanoscale (invisible) machines and robots, also referred to as "*machine-phase nanotechnology*." The *reality* definition of nanotechnology is a synonym for fundamental materials and matter research, including quantum phenomena at small length scales. The origins are clear.

For the past 40 years, since the invention of the semiconductor transistor, the economic apparatus built around *Moore's Law* has been driving material features ever smaller. Concurrently, our understanding of the basic quantum physics that governs the behaviour of interacting particles (of matter and light) has been solidifying. More recently, measurement and fabrication techniques have reached a point where we can start thinking seriously about exploiting some of these novel properties that appear in the small length regime. As this level of control gets closer technologically, with clear opportunities in sight, the funding of these endeavors becomes more worthwhile. If nanotechnology can act as an umbrella term to drive interest and funding, so be it. The head of the NNI publicly espouses this viewpoint.

We can further separate the larger field nanotechnology, in terms of the reality and vision accompanying this emerging technology, from the recent interest in nanoparticle and quantum dot technologies. In many ways, certainly as far as environmental and human toxicity are concerned, nanotechnology can be defined much more narrowly than the above: Nanotechnology, at present, is *nanoparticles* and *nanomaterials* that contain nanoparticles. *Nanoparticles* are defined as objects or devices with at least two dimensions in the nanoscale regime (typically tens of nanometers or less) that exhibit new properties, physical, chemical, or biological, or change the properties of a bulk material, due to their size. Nanotechnology of the future will include atom-by-atom or molecule-by-molecule built active devices.

Much of the excitement surrounding nanotechnology comes from the promise of newly gained nanoparticle synthesis techniques and a realization of their potential in many different areas. Two prominent examples are bio-markers for cancer detection (and destruction) and quantum dots in solar-energy conversion devices. Nanoparticle and nanoparticle-composite technology may end up solely a minor or micro revolution separate from the broader nanotechnology field

Nanotechnology's Place in an Age of Ages

It is not only the speed of technological change that creates a "revolution," it is its scope as well. Above all, today, as seven thousand years ago, technological developments from a great many areas are growing together *to create a new human environment.*

Drucker's words remain true. Nanotechnology as presently (loosely) defined will likely have several acts to play in the coming century. Certainly the utilization of nanoparticle technology has immediate promise. Much of the rest of nanotechnology in the near term can more accurately be placed within the information or biotechnology revolutions. In either case, we can find patterns. The five major revolutions that Perez outlined all follow a similar pattern. The first stage is the installation period, which has an eruption phase, when a new innovation is introduced and spreads in conflict with old products and technologies. The second is the frenzy phase, when financial capital drives the build-up of new technologies but develops tensions within the system.

A turning point occurs, usually with a recession that follows the collapse of a financial bubble, and regulatory changes are made to facilitate and shape the period of development. Then follows a period of deployment, which initially has a synergy phase, when conditions are all favourable for the full flourishing of the new technology, and then the maturity phase, when signs of dwindling investment opportunities and stagnating markets appear. Obviously there is much fluctuation in this model. Since nanotechnology as labeled takes on so many meanings, we must separate the key components. First, there is the nanoparticle/quantum dot component, and we will call this the nanoparticle revolution. Second, there is a continuation of technologies resulting in nanoscale techniques for manufacture that are being widely adopted by big industry (GE, Dupont, Intel). We can hesitantly call this a second industrial revolution (ground up technology?).

Finally, in the far distance, there is the machine-phase nanotechnology revolution, completely imaginary at this stage. Only this stage of development (promising essentially free goods) holds the potential for drastic social and political upheaval. At the current state of development, nanotechnology simply does not represent a paradigm shift in scientists' thinking. Nanoscale investigation is an evolutionary outgrowth of a new capability to measure and fabricate at that scale. Nanotechnology must be seen in the greater trend of innovation, which, like the irrigation revolution, will likely continue well into the next millennium.

Military and Technological Development

The military has long been an instigator and shaper of technological innovation. Often the high-cost buyer or buyer of last resort, the military can act both to encourage a fledgling technology and to prolong a dying one. The US *Department of Defense* (DOD) has clearly taken an interest in nanotechnology and accounts for roughly 28% of all federal funding in the loosely defined field in FY2005. One prominent example is the Institute for Soldier Nanotechnologies at the Massachusetts Institute of Technology. Stronger and lighter materials and more explosive bombs (*super thermites*) are but two examples of nanotechnology's impact on future warfare.

It is widely assumed that military-born technologies spill over into civilian use for beneficial purposes. However, the military has specific objectives in its approach to technology, which might be very different from those of society at large. Historian David Noble lays out three such objectives in his treatise on military and technology (Noble, 1987) which are worth considering again in the context of nanotechnology. These are (i) performance (emphasis placed on meeting military objectives and what follows necessarily from them), (ii) command (management techniques with decision-making coming solely from the top), and (iii) modern methods (a fetish for machinery that won't talk back). Noble argues that it is a misconception that the military acts only as an external input of technology. Instead, the military shapes the progress and nature of a technology or set of technologies throughout their lifetime in many cases.

One clear example of this dates back to the beginning of the US as a nation where the military's quest for interchangeable gun parts helped spur mechanization and the industrial revolution in the States. Uniformity was imposed by the military contract system. "The benefits of the system, clear to the military, were not so clear to many manufacturers, given the high costs, uncertainties, and inescapable industrial conflict it engendered" (Noble, 1987). A similar example is that of *numerical control*, pioneered by the Air Force in the 1980s. Numerical control envisioned extremely precise machining based on computer and mathematical specification and extreme shortening of the chain of command from aircraft part specification to manufacture. Industry generally was not enthusiastic as the systems were very complex and not as flexible as other less-demanding, though adequate, metal working techniques. Since the military provided such a large and stable base of funding, however, industry followed the numerical control path (Noble, 1987). Industry paid the price as foreign competition became more nimble. The loss of promising alternative technologies, excessive consolidation in the metalworking industry, and slow innovation all resulted from the military's involvement.

An excellent counter-example to this phenomenon is Intel. Although the military was the initial buyer of Intel's first few-transistor circuits, its preferences did not shape Intel's future. Intel would not have survived in the rapidly changing consumer environment had its engineers been unable to make decisions. In fact, Intel has thrived on a very long chain of command. In other words, engineers very near the technology (but at the bottom of the corporate hierarchy) are entrusted with a large amount of discretion to make decisions related to technology undeniably vital to the company's future. The rate of growth in the private sector made this possible, although it is important to note that some semiconductor fabs continued at a reduced level by specializing their wares for military needs (think Fairchild Semiconductor, a founding company of silicon valley which has since largely left the commercial consumer electronics arena for mostly military and advanced technology contracting).

The long-term trend is an increasing shift of federal research dollars into the mission-driven agencies and away from discipline-driven research, such as in NSF, the *Department of Energy's* (DOE) Office of Science, and at NIST (*National Institute of Standards and Technology*). Apart from the *National Institutes of Health* (NIH), non-defense federal R&D is about the same in 2004 dollars as it was in 1980. Basic, unfettered physical research in the US is declining, except where it goes through the DOD mission agencies and the NNI.

In general, the military funding agencies are much stricter about how their grant money is used as compared to the NSF. Since all or most of the research money comes through the military, and they are the ones asking hard questions and threatening to pull funding, scientists at universities feel strong pressure to follow the dictated "*roadmaps*" instead of pursuing new physics as it is identified. Because of this, new and perhaps useful phenomena at the nanoscale - which may lay the groundwork for the next revolution 50 years hence - may be missed in this country.

Lessons from the Past

A recent commercial by *General Electric* featured a "professor of nanotechnology" and a super model falling in love: "the perfect combination of brains and beauty" (GE, 2005) "*Nanotechnologist*" has become the new computer scientist, driver of the next great wealth generator. While easy to dismiss, it is important to remember that but for the abnormal obsession of a couple dozen people with the properties of semiconductors, the US would not have led the personal computing, networking, and internet revolutions of the last half of the 20th Century. A case study is Great Britain, who irreversibly fell behind Germany and the US because it faltered in its investment in new technologies during the age of steel, electricity, and heavy engineering (1875-1920). Will history repeat itself in the US?

Before World War II, US universities were a joke internationally. Due to the demonstrable success of radar and the atomic bomb, the US quickly realized that science played a key role in military victories and national power, so a large-scale investment in fundamental research began. The GI Bill supplied manpower. Through this and America's survival as a superpower after the war (and a concurrent influx of highly trained European scientists), the US has had the good fortune to lead the last major techno-economic revolution: the age of information and telecommunications, as well as many minor ones. But as other countries catch up to America's core strengths, the US leadership position is tenuous. Indeed, funding in nanotechnology in Japan and Europe is comparable to that of the US at present.

If we want to lay the foundation for the next revolution, it is instructive to go back and try to understand what began in the mid 18th century in England. In fact, the name Industrial Revolution is a misnomer, as innovations took place in many areas such as farm and home, in addition to manufacturing. No one knows for sure why the Industrial Revolution began where and when it did. There are many hypotheses: Britain's institutional support of technology (world's first patent system, strong private property rights, acceptance of Jewish and other ethnic minorities); urbanization and increased life expectancy; encouragement of an empirical and utilitarian tradition; consolidation of agricultural land by lords with agricultural efficiencies; increased worker migration to cities; movement of work away from the guild system (putting out system); raw material advantages; less regulation. The list goes on. But other European countries like Germany and France, who also had better educational systems, shared many of these advances in whole or in part.

The one thing other countries lacked was a transport system even remotely comparable to what England had put in place. "Transport improvements greatly accelerated the processes of regional specialization and urbanization in England. They also led to a dramatic increase in personal travel". This encouraged the interaction between innovators with varied backgrounds, expertise, and ideas, which is essential to the innovation process. Regionalization and localization led to mechanization. In the present day, the US has a mixed infrastructure in idea transportation. Although it has pioneered advancements in collaboration and interaction on-line, several other countries, such as South Korea or Taiwan have superior broadband networking penetration. Residents of the US have always enjoyed freedom to move about the country, and career success often demands it. The US also enjoys the benefits of scale, with a large number of excellent yet independent universities and a large entrepreneurial culture (exhibited in individuals and in organizations such as top-notch private-equity entities).

Presently, first-world countries like Great Britain, Australia, and Japan - although investing heavily in nanotechnology research - are struggling to match the US's highly efficient venture capital ecosystem. However, immigration rules since 9/11 have decreased the influx of talent from around the world, traditionally a key driver of science research. But none of this compares to the great experiment that is ongoing in the very nature of science investment in the US. Industrial science and technology in the US has undergone a dramatic change in recent years, from "Closed Innovation" to "Open Innovation".

In essence, the era of industrial research labs is over. Where significant basic research used to occur in the bowls of Bell Labs or Xerox PARC, industry has now focused more on development of near and more economically justifiable engineering. Extreme examples of this are companies like Intel and Cisco, who "*outsource*" virtually all their research. They leverage their research budgets by partnering with academia and other companies and start-ups. This is different from and in addition to what's usually called outsourcing – the farming out of actual work or jobs (in this case in research and development) to countries such as China and India. "Under Open Innovation, a company's value chain is no longer fully contained within the company, and ideas, people, and products flow across company boundaries, to and from other companies, universities, and even countries".

This business trend has left only universities and national labs to fulfill the need for basic research in the US. From 1953 to 1996, the fraction of basic research that was performed in universities and federal labs rose from 33% to 61%. Is this enough to make up the difference? We simply do not yet know how this change will affect US competitiveness in the future. "The growth of biotechnology in America is largely a story of seedling ideas that came from academic scientists in research universities, funded by venture capitalists, and manned by bright graduate and postdoctoral students". Nanotechnology may prove to be the same story, or not.

There are a number of conditions that allowed the Industrial Revolution to move quickly to the US: fast population growth; natural and artificial protection (via the Atlantic Ocean and tariffs); copying and extending prior work (of the British banking system, corporate, and insurance; manufacturing techniques, etc.); relief from bankruptcy (limited liability); legal monopoly over inventions through patent law – with strict granting of patent applications to ensure that only new and useful ideas were patented; lack of guild monopolies; vast natural resources; receptivity to innovation. What country today has the most of these benefits? As Duke and Dill point out, the US must focus on its core strengths: innovative and fast-moving companies, talented people, and strength in basic research. With these concerns, the motivation of the NNI to pump money into fundamental research under the cover of nanotechnology seems like a very good move. While on the surface the business trend to open innovation seems a good way to speed up business and technology growth, it is unclear what the long term affects on the US will be. If the majority of basic research ends up in Asia, can US corporations seriously believe they will be allowed to "manage" and benefit from these new discoveries indefinitely?

That we are living through a great technological revolution with no end in sight is clear. What gets murky is our attempt to subclassify smaller revolutions within this larger landscape of merging innovations. The reality of nanotechnology as it stands today is the continued evolution of prior trends in information and materials science began after World War II. That we are at a point where the exact synthesis of nanoparticles and other nanotechnologies holds great promise for medicine, energy conversion, etc., has only fueled the belief that a renewed surge of investment is needed to harvest these potential technological breakthroughs. We have neither begun to approach the vision of nano-machines and robots that popularized the term nanotechnology, nor to adequately understand the difficulty in getting there. So the great technological revolution that this may imply lies still in waiting for us to discover.

Nanofuture

The history of disease is vastly older than that of humankind itself. Indeed, disease and parasitism have been inseparable companions to life since the dawn of life on Earth. Fossilized bacteria similar to those responsible for many infections that afflict people today have been found in geological formations that are 500 million years old. Fossil shells dating from an era almost equally remote show clear evidence of disturbance by injury and parasites. Examination of the skeletons of long-extinct dinosaurs and other great reptiles show that these creatures suffered from fractures, bone tumours, arthritis, osteomyelitis, dental caries and other diseases that still plague us in the 20th century. While available fossil evidence is largely limited to changes observable in bones and teeth, it is probably safe to assume that disease processes were equally prevalent in the soft organs and tissues that have not been geologically preserved, and that the general pattern of disease has not changed in its essentials during the hundreds of millions of years that animal life has existed on this planet.

Since its first appearance on the prehistoric stage millions of years ago, the human body has also been constantly subject to assault and injury, invasion by parasites, extremes of heat and cold, and infections. Early man probably suffered from a number of diseases due to nutritional factors and body chemistry disorders. For example, the poor condition of the teeth of an 18-year old Australopithecus who lived 1.75 million years ago, found at Olduvai Gorge by Louis Leakey in 1959, suggests that the hominid had a disease that lasted for many months—most likely gastro-enteritis due to malnutrition—with three major attacks of the disease at the ages of two, four, and four and a half. Diseases that may date back more than 25 million years to the ape ancestors of modern apes and man are amoebic dysentery, malaria, pinworm infections, syphilis, yaws, and yellow fever. Diseases which may have appeared and evolved with man include leprosy and typhoid; certain modern diseases such as cholera, measles, mumps, smallpox, whooping cough, and the common cold require large concentrated populations to support them, thus probably could not have existed in the prehistoric era.

As for cancer, Java man, first discovered by Dutch anatomist Eugene Dubois in 1891 and considered to be half a million years old, had a morbid growth on his femur. It is likely that bone cancer and other forms of cancer have existed from the earliest times. The remains of *Neanderthal*, a competing species to *Homo sapiens* that roamed through Europe, Africa, and the Near East during the last glacial period ~75,000 years ago, show clear evidence of arthritis, tooth loss, and suppurative bone disease. (The high rate of broken bones and early death suggests that *Neanderthals* engaged in more close-quarter combat with large animals than did modern humans, who had figured out safer strategies.) Human bones unearthed from the New Stone Age period reveal that Neolithic man suffered from arthritis, congenital dislocations and fractures, sinusitis, tuberculosis of the spine, and tumours.

In considering the question of how our earliest ancestors dealt with these conditions, we are on somewhat uncertain ground, since little direct evidence has been preserved. Anthropologists point out that in pre-*Neanderthal* hunter-gatherer tribes, a sick or lame person is a serious handicap to a group on the move. In the event of major illness or mortal wounds, sufferers may either leave the group, be abandoned, or, as with lepers in medieval Europe, may be ritually expelled, becoming "culturally dead" before they are biologically dead. Early hunter-gatherer hominid bands were more likely to abandon their seriously sick than to succor them, although *Cro-Magnons* and *Neanderthals* evidently were the first to care for their wounded and disabled, and to bury their dead.

But moderate wounds, bruises, fractures, or foreign bodies such as arrowheads or thorns are tangible things that demand attention. Thus the art of surgery must first have originated as a response to immediate crises. The first and most obvious course of action of a wounded man would be to protect the site of injury from the influence of external forces or agents. For this there was, and remains to this day, only one means—the application of a dressing. Many observations were made and many substances tried, and in time a body of experience was accumulated and passed on orally to others for use in similar emergencies, eventually creating a considerable sum of inherited empirical knowledge. It is believed that the art of dressing wounds long constituted the whole of medicine—the use of internal remedies or herbs, and use of the knife or fire, came much later.

Present-day primitive and folk medicines provide additional clues to early medical practice. For example, many early peoples developed effective methods to control bleeding—the use of cobwebs is an ancient folk remedy, as is the application of tourniquets, packing with absorbent materials, the laying on of snow or, at the other temperature extreme, the application of cautery by hot knife or spear. The *Masai* and *Akamba* tribes treated sword wounds by slapping on a poultice of cow dung and dust. The suturing of wounds is practiced by some primitive peoples and may have been known to prehistoric man. Bone needles furnished with an eye have been found in paleolithic deposits in France and England, and some Indian tribes suture with threads of sinew or bone needles (the needles are left in and the thread is twisted around them). One of the strangest suturing techniques, observed among primitive tribal cultures in such widely separated places as India, East Africa and Brazil, is the sealing of wounds, especially abdominal wounds, using termites or ants. The edges of the wound are drawn closely together and the insect is allowed to bite through them both, firmly securing the flesh on two sides. Once attached, the insect's body is severed, allowing only the jaws to remain in place, holding the wound shut.

While able to deal with wounds, early man did not admit the existence of disease from "natural causes". Internal diseases were generally ascribed to malevolent influences exercised by a supernatural entity or a human enemy. As centuries passed, many internal diseases were eventually recognized and simple treatments empirically established—for example, Celsus (ca. 30 AD), a Roman medical writer, described ligature (tying off blood vessels), suturing the large intestine, eye operations such as couching for cataract, tonsillectomy, and bladder surgery to remove stones. But the rational basis for internal medicine awaited basic physiological knowledge such as the circulation of the blood, finally proven by William Harvey (1578-1657) in 1628, and the discovery and acceptance of the theory of infection by microorganisms in the 1800s. Throughout most of human history, life has been short indeed. Even by the 17th century, more than half of all children never lived past the age of ten, often succumbing to diseases such as cholera, diphtheria, scarlet fever, or whooping cough, or being scarred for life by smallpox. It was a common and widely accepted fact of life that people would die at all ages, although wealthy families could leave town every year during the cholera season. One of the constant missions of human civilization has been the avoidance and elimination of animals that prey on humans. Cave-dwelling carnivorous saber-toothed tigers, having occupied the upper echelons of the food chain for 30-35 million years, finally became extinct not more than about ten thousand years ago, most likely at the hands of newly-arrived human hunters crossing the Siberian land bridge into post-Pleistocene North America.

As late as medieval times, wolves ranged freely over Europe, remaining abundant in France through 1500 AD—in winter, audacious wolf packs would enter Paris and eat children, dogs, and even adults who were alone on the streets. With technological advances, extant tiger and wolf species are now largely confined to artificial habitats or isolated nature preserves and no longer pose any serious threat to human health. People in most places are not eaten by wolves; indeed today, a few venturesome individuals actually keep and breed wolves with dogs, as pets. Bacteria are among the last remaining "wild animals" on Earth that threaten man. As our instrumentalities continue to progress from the macroscale to the microscale, and finally to the molecular or nanoscale, all of the remaining natural "wild things" that endanger human life and health—whether viruses, bacteria, protozoa, metazoan parasites, or even our own pathological native cells—will be confined and tamed, reconstructed or eliminated. Using smaller tools, we hunt smaller prey. As with sabertooths in the post-Neolithic era, and with wolves in post-medieval times, people of the 21st century will no longer fear or need suffer predation by wild microbes or tumour cells run amok.

Humanity is poised at the brink of completion of one of its greatest and most noble enterprises. Early in the 21st century, our growing abilities to swiftly repair most traumatic physical injuries, eliminate pathogens, and alleviate suffering using molecular tools will begin to coalesce in a new medical paradigm called *nanomedicine*. Nanomedicine may be broadly defined as the comprehensive monitoring, control, construction, repair, defense, and improvement of all human biological systems, working from the molecular level, using engineered nanodevices and nanostructures.

Current Medical Practice

In order to fully appreciate the changes that nanomedicine will inevitably bring, it is useful first to review the history and development of current medical practice. As Winston Churchill once remarked: "The further backward you look, the further forward you can see." The author unapologetically favours what medical anthropologists would regard as the "Western" healing tradition. The great medical theorist Otto E. Guttentag agreed:"Contemporary Western medicine involves a type of healing that potentially and actually exceeds all other approaches in maintaining that optimal status of selfhood we call being healthy and in eliminating that reduced status of selfhood we call being sick."

Evolution of Scientific Medicine

A study of the history of "scientific" or Western medical practice suggests a continuity in certain aspects of the medical paradigm even since ancient times—as for instance the basic precepts of observation and diagnosis, followed by treatment—but reveals changes in attitudes, techniques, and instrumentalities, and a gradual evolution in the epistemology and ontology of medical logic. Note that 2010 is marked as a possible date for the first applications of *nanomedicine*. These will most likely be *ex vivo* applications only *in vivo* nanomedical treatments may come much later.

Another important reason to study the history of medicine is to gain a deeper appreciation of the long, hard struggle to improve human health, a struggle that is expected finally culminate in victory in the 21st century. If the ~10 billion people that have ever lived survived an average of 40 years and spent 5% of their lives in sickness or physical misery from disease, then ~200 trillion man-hours of suffering have been paid to achieve this remarkable result, a not inconsiderable price. There is no pretense to completeness here. For example, Chinese, Indian and Islamic contributions are omitted, not because they are unimportant, but because they did not significantly alter the evolutionary pathway of the Western medical paradigm. A great deal of physiology, pathology, neurology, systematics, and many other important medical-related disciplines are also neglected in the interest of brevity.

Prehistoric medicine

The elementary nature of prehistorical medical practice has already been mentioned. There is direct evidence that *Stone Age* man used a natural fungus as a treatment for intestinal parasites. Besides basic wound-tending, the practice of circumcision is an age-old form of simple surgery with a rational, if sometimes controversial, hygienic basis.

An even more dramatic surgical operation, for which there is considerable prehistoric fossil evidence, is trephining the skull to remove a round piece of bone from it. This can be done using flint instruments, either by gradually scratching through the skin and bones of the skull, getting gradually deeper and deeper, or by drilling a series of small holes in a circle in the skull, then cutting the small bridges between to remove a disk of bone. Skulls have been found with multiple holes. The additional fact that some hole edges show callus formation (evidence of healing) indicates survival of the patient. Trephined skulls have been found in Western Europe, including England, North Africa, Asia, the East Indies, New Zealand, and the Americas from Alaska in the north down through the continent to Peru in the south. The practice was probably regarded as therapeutic, either to remove a depressed fracture, to try to cure mental illness, or to relieve severe headache or epilepsy, presumably by letting out the demon possessing the patient.

Ancient mesopotamian medicine

The first written records, which came from ancient Babylon and Egypt, contain the earliest references to medical care,obviously codifying earlier practices no record of which has survived. The medicine of the Sumerians of Mesopotamia from ca. 3000 BC was primarily religious. The Mesopotamian peoples saw the hands of the gods in everything. Disease was caused by spirit invasion, sorcery, malice, or the breaking of taboos, and sickness was both judgement and punishment. An Assyrian text circa 650 BC describes epileptic symptoms within a demonological framework: "If at the time of the possession, his mind is awake, the demon can be driven out; if at the time of his possession his mind is not so aware, the demon cannot be driven out." Headaches, neck pain, intestinal ailments and impotence were read as omens. The appropriate remedy was to identify the demons responsible and expel them by spells or incantations.

But medicine also had an empirical component, with some sicknesses being ascribed to cold, dust and dryness, putrefaction, malnutrition, venereal infection and other natural causes. The Babylonians drew on an extensive pragmatic materia medica—some 120 mineral drugs and twice that number of vegetable items are listed in surviving tablets. Alongside various fats, oils, honey, wax, and milk, were many active ingredients that included mustard, oleander and hellebore (a plant in the buttercup family that is a violent gastro-intestinal poison, hence acts as a powerful purgative, though it is lethal in high doses). Colocynth, senna and castor oil were used as laxatives, while wound dressings were compounded with dried wine dregs, salt, oil, beer, juniper, mud or fat, blended with alkali and herbs. With the discovery of distillation, the Mesopotamians made essence of cedar and other volatile oils. Turpentine, asafetida, henbane, myrrh, mint, poppy, fig, and mandrake are also mentioned. Dog dung and other fecal ingredients were used to drive off demons.

Surgical conditions such as wounds, fractures and abscesses were also treated by Mesopotamian surgeons. Practitioners were priests, and after 2000 BC, they were ruled by the strict laws included in the Code of Hammurabi. The Code laid down rewards for success and severe punishment for failure, and contained laws relating to medical practice which show that medicine and surgery were highly organized professions. Fees were regulated on a sliding scale of rewards based on the patient's rank, and severe penalties were laid down for failure:

"Concerning the wounds resulting from operations it is written: if a physician shall produce on anyone a severe wound with a bronze operating knife and cure him, or if he shall open an abscess with the operating knife and preserve the eye of the patient, he usually shall receive 10 shekels of silver [more than a craftsman's annual pay]; if it is a slave, his master shall usually pay 2 shekels of silver to the physician."

"If a physician shall make a severe wound with an operating knife and kill him, or shall open an abscess with an operating knife and destroy the eye, his hands shall be cut off."

"If a physician shall make a severe wound with a bronze operating knife on the slave of a free man and kill him, he shall replace the slave with another slave. If he shall open an abscess with a bronze operating knife and destroy the eye, he shall pay the half of the value of the slave."

The Hammurabic Code also mentions the Gallabu, or barber-surgeons, whose province was minor surgery, including dentistry and the branding of slaves. If Herodotus (ca. 485-425 BC) may be believed, Babylonian medicine must have declined in the 5th century BC. Herodotus states that there were no physicians, but that the people brought their sick into the marketplace in order that passers-by might make suggestions or offer cures.

Ancient Egyptian medicine

The civilization of ancient Egypt dates from around 3000 BC. As in Mesopotamia, ancient Egyptian medicine was religious-empirical. An examination of the preserved bodies of members of the Royal family has provided much information on the diseases of ancient Egypt, including congenital deformities such as clubfoot, dental caries, gallstones, bladder and kidney stones, rheumatoidarthritis, mastoiditis, numerous eye diseases, and bone fractures, some of which show treatment by quite sophisticated splinting. Other evidence suggests that the average ancient Egyptian was extremely diseased. For example, a weaver who died in the 11th century BC, aged 14-18 years, evidently suffered from *schistosomiasis*, tapeworm likely associated with malnutrition, *anthracosis* of the lungs presumably due to environmental pollution from cooking and heating, *pulmonary silicosis*, and possibly malaria and fleas.

The Ebers Papyrus (ca. 1550 BC), deriving from Thebes, is the principal medical document and may be the oldest surviving medical book. Over 20 meters long, it deals with scores of diseases and proposes remedies involving spells and incantations, but also includes many rational treatments. The Ebers Papyrus covers 15 diseases of the abdomen, 29 of the eyes, and 18 of the skin, and, perhaps unsurprisingly to the modern consumer, lists no fewer than 21 cough treatments. About 700 drugs and 800 formulations are mentioned, mainly involving herbs but also including mineral and animal remedies. For example, to cure night blindness the patient should eat fried ox liver—possibly a tried-and-tested procedure, since liver is rich in Vitamin A, lack of which causes the illness. Eye disorders were also common, for example:

"To drive away inflammation of the eyes, grind the stems of the juniper of Byblos, steep them in water, apply to the eyes of the sick person and he will be quickly cured. To cure granulations of the eye, prepare a remedy of cyllyrium, verdigris, onions, blue vitriol, powdered wood, and mix and apply to the eyes."

For stomach ailments, a decoction of cumin, goose-fat and milk was recommended, but other remedies sound more exotic, including a drink prepared from black ass testicles. A mixture of vulva and penis extracts and black lizard was supposed to cure baldness. Also good for hair growth was a compound of hippopotamus, lion, crocodile, goose, snake and ibex fat—merely assembling these ingredients might promise a hair-raising experience! Egyptian medicine credited many vegetables and fruits with healing properties, and also tree products such as sycamore bark and resins such as myrrh, frankincense and manna. As in Mesopotamia, plant extracts—notably senna, colocynth and castor oil—were employed as purgatives, and oil of camomile to improve digestion. Recipes included ox spleen, pig's brain, stag's horn, honey-sweetened tortoise gall, and the blood and fats of various animals. Antimony, copper salts, alum, and other minerals were recommended as astringents or disinfectants. Containing ingredients from leeks to lapis lazuli—including garlic, onion, tamarisk, cereals, spices, condiments, resins, gums, dates, hellebore, opium and cannabis—compound drugs were administered in the form of pills, ointments, poultices, fumigations, inhalations, gargles and suppositories; they might even be blown into the urethra through a tube. Among the most interesting of the healers whose names have been recorded for posterity are Peseshet, head female physician or overseer, proof of the existence, as in Mesopotamia, of female healers; and Iri, Keeper of the Royal Rectum, presumably the pharaoh's enema expert.

The Edwin Smith Papyrus (ca. 1600 BC), discovered by its American namesake at Luxor in 1862, may be the world's earliest surviving surgical text, and contains material probably derived from even more ancient times. The "*book of wounds*" comprises 48 case reports, which commence with the top of the head and proceed systematically downward —nose, face, ears, neck and chest, mysteriously stopping in mid-sentence at the spine, presumably when the scribe was interrupted at his work. The only surgical conditions treated were wounds, fractures, abscesses, and circumcisions. In the Papyrus, the method of presentation is first to set out the title or the chief symptom, followed by the further symptoms, and then the examination, diagnosis, prognosis, and treatment. The advice given is entirely rational, as will be seen by the following directions for the treatment of a fractured humerus:

"Instructions concerning a Break in his Upper Arm. If thou examinest a man having a break in his upper arm, and thou findest his upper arm hanging down, separated from its fellow, thou shouldst say concerning him: One having a break in his upper arm. An ailment which I will treat."

"Thou shouldst place him prostrate on his back, with something folded between his two shoulder blades; thou shouldst spread out with his two shoulders in order to stretch apart his upper arm until that break falls into its place. Thou shouldst make for him two splints of linen, and thou shouldst apply for him one of them both on the inside of his arm, and the other of them both on the underside of his arm. Thou shouldst bind it with ymrw (an unidentified mineral substance), and treat it afterward with honey every day until he recovers."

An interesting point in the case histories is that after the diagnosis the writer gives a decision about his further course of action. The verdict may take one of three forms: (i) an ailment which I will treat; (ii) an ailment with which I will contend; and (iii) an ailment not to be treated. Like modern military triage, the "hopeless" patient is left to his inevitable fate. This guarded attitude on the part of the medical man was widespread in antiquity. While present-day doctors generally do everything possible to alleviate symptoms to the very end, even when the patient has no chance of recovery, the view in ancient times was that hopeless cases were not to be touched. This attitude was entirely practical. A doctor in attendance at the courts of ancient Egypt might expect rich rewards if his patient recovered, but if a patient died under his care, the unfortunate physician ran a grave risk of impalement.

Ancient Greek medicine

By 1000 BC the communities later collectively known as the Greeks were emerging around the Aegean Sea. How much medical knowledge they took from Egypt remains controversial, but the contrasts between the two are striking. Little is known of Greek medicine before the appearance of written texts in the 5th century BC. Archaic Greece had its folk healers, including priest healers employing divination and herbs. From early times (the first Olympic games were recorded in 776 BC), the love of athletics produced instructors in exercise, bathing, massage, gymnastics and diet. The Homeric epics (ca. 600 BC) offer glimpses of early Greek medicine. Scholars count 147 cases of battle wounds in the Iliad, including 106 spear thrusts, 17 sword slashes, 12 arrow shots, and 12 sling shots. Among the arrow wound survivors was King Menelaus of Sparta, whose physician extracted the arrow, sucked out the blood and applied a salve. As with other medical interventions in Homer, this shows no Egyptian influence, supporting the idea that, even if Greek practice owed much to Egypt, it rapidly went its own way.

Various Greek gods and heroes were identified with health and disease, the chief being *Asclepius*, who even had the power to raise the dead. A heroic warrior and blameless physician, Asclepius was the son of Apollo, sired upon a mortal mother, who was taught herbal remedies by Chiron and then generously used them to heal humans. Incensed at being cheated of death, Hades, the ruler of the underworld, appealed to the supreme god, Zeus, who obligingly dispatched Asclepius with a thunderbolt, though he was later elevated to godhood. A different version appears in Homer, who portrays Asclepius as a tribal chief and a skilled wound healer whose sons became physicians and were called Asclepiads, and from whom all Asclepian practitioners descended. As the tutelary god of medicine, Asclepius is usually portrayed with a beard, staff and snake—the origin of the caduceus symbol of the modern physician, with its two snakes intertwined, double-helix like, on a winged staff. The god was often shown accompanied by his daughters, Hygeia (health or hygiene) and Panacea (cure-all).

For all that, *Hippocratic medicine*, the foundation of Greek written medicine, explicitly grounds the art upon a quite different basis—a healing system independent of the supernatural and built upon natural philosophy. The beginnings of true medical science in the West were established when the reliance on superstition that underpinned tribal medicine was replaced by civilized and rational curiosity about the cause of illness. The growth of civilized thought allowed for argument on medical cause and cure, with great doctrinal multiplicity. The separation of medicine from religion reveals another distinctive feature of Greek healing: its openness, a quality characteristic of Greek intellectual activity in general, owing to political diversity. There was no imperial Hammurabic Code and, unlike *Egypt*, no state medical bureaucracy, nor were there examinations or professional qualifications. Those calling themselves doctors (iatroi) had to compete with bone-setters, exorcists, root-cutters, incantatory priests, gymnasts, and showmen, exposed to the quips of playwrights and the criticism of philosophers. Medicine was open to all.

Empedocles (fl. 450 BC) may have been the first to advance some of the key physiological doctrines in Greek medicine, including innate heat as the source of living processes such as digestion, the cooling function of breathing, and the notion that the liver makes the blood that nourishes the tissues. His contemporary, *Alcmaeon of Croton* (fl. 470 BC), believed that the brain, not the heart, was the chief organ of sensation. Alcmaeon's examination of the eyeball led him to discern the optic nerve leading into the skull, a genuine observational basis. He gave similar explanations for the sensations of hearing and smelling, because the ear and nostrils suggested passages leading to the brain. Most of such knowledge depended heavily on wound observation and animal dissection, for in the classical period the dignity of the human body forbade dissection.

All we know for sure of *Hippocrates* (ca. 460-377 BC), who "taught all that were prepared to pay," is that he was born on the island of Cos and lived a long and virtuous life. The sixty or so works comprising the *Hippocratic Corpus* (ca. 440-340 BC) derive from a variety of hands, and, as with the books of the Bible, they became jumbled up, fragmented, then pasted together again in antiquity. What is now called the *Corpus* was gathered around 250 BC in the Library at Alexandria, with further texts added later still. Some volumes are philosophical, others are teaching texts or case notes. What unites them all is the conviction that health and disease are capable of explanation by reasoning about nature, independently of supernatural interference. Man is governed by the same physical laws as the cosmos, hence medicine must be an understanding, empirical and rational, of the workings of the body in its natural environment. Anticipating modern medicine, appeal to reason, rather than to rules or to supernatural forces, gives Hippocratic medicine its distinctiveness. It was also patient- rather than disease-centered; the Hippocratics specialized in medicine by the bedside, prizing trust-based clinical relations: "Make frequent visits; be especially careful in your examinations, counteracting the things wherein you have been deceived at the changes. Thus you will know the case more easily, and at the same time you will also be more at your ease. For instability is characteristic of the humours, and so they may also be easily altered by nature and by chance."

"...Keep a watch also on the faults of the patients, which often make them lie about the taking of things prescribed. For through not taking disagreeable drinks, purgative or other, they sometimes die. What they have done never results in a confession, but the blame is thrown upon the physician."

The Hippocratics promulgated the idea of "*vis medicatrix naturae*," or the power of nature to cure itself, and thus the belief that there was a natural tendency for things to get better on their own. This tendency could be aided by providing a beneficial environment for the patient and by improving physical function with a regimen of suitable diet, lifestyle, and exercise—the diatetica. In extreme cases, further aids to recovery could be sought. Stubbornly offending "*humors*" could be removed with the help of venesection (phlebotomy or bloodletting) and purgatives, sudorifics applied to induce sweating, and diuretics to increase urination. But the Hippocratic physician was extremely reluctant to administer drugs of any kind, because the physician's goal was to aid nature in healing the body.

Hippocratics scorned heroic interventions and left risky procedures to others. Their Oath explicitly forbade cutting, even for stones, and other texts reserved surgery for those used to handling war wounds. Surgery was regarded as an inferior trade, the work of the hand rather than the head, a fact reflected in its name: "surgery" derives from the Latin "*chirurgia*," which comes from the Greek "cheiros" (hand) and "ergon" (work); surgery was handiwork. Hippocratic surgical texts were thus conservative in outlook, encouraging a tradition in which doctors sought to treat complaints first through management, occasionally through drugs, and finally, if need be, by surgical intervention.

The art of diagnosis involved creating a profile of the patient's lifestyle, work and dietary habits, partly by asking questions and partly by the use of trained senses: "When you examine the patient, inquire into all particulars; first how the head is...then examine if the hypochondrium [abdomen beneath lower ribs] and sides be free of pain, for...if there be pain in the side, and along with the pain either cough, tormina [painful intestinal colic] or bellyache, the bowels should be opened with clysters [enema].... The Physician should ascertain whether the patient be apt to faint when he is raised up, and whether his breathing is free..."

Hippocratics prided themselves on their clinical acuity, being quick to pick up telltale symptoms, as with the facies hippocratica, the facial look of those dying from long-continued illness or cholera: "a protrusive nose, hollow eyes, sunken temples, cold ears that are drawn in with the lobes turned outward, the forehead's skin rough and tense like parchment, and the whole face greenish or black or blue-grey or leaden." Experience was condensed into aphorisms, as for instance: "When sleep puts an end to delirium, it is a good sign."

The technique most prized among Hippocratics was the art of prognosis—a secular version of the priestly and oracular prognostications of earlier medicine, and bearing some analogy to the 20th century weatherman, who can give a bright or gloomy forecast but is powerless to change it. Noted one Hippocratic text: "It appears to me a most excellent thing for the physician to cultivate Prognosis; for by foreseeing and foretelling, in the presence of the sick, the present, the past, and the future, and explaining the omissions which patients have been guilty of, he will be the more readily believed to be acquainted with the circumstances of the sick; so that men will have confidence to entrust themselves to such a physician....Thus a man will be the more esteemed to be a good physician...from having long anticipated everything; and by seeing and announcing beforehand those who will live and those who will die."

The ultimate significance of Hippocratic medicine was twofold. First, it carved out a lofty role for the selfless physician which would serve as a lasting model for professional identity and conduct. Second, it taught that an understanding of sickness required an understanding of nature.

Ancient Alexandrian medicine

Soon after the death of *Aristotle* (d. 322 BC) and his most famous pupil, *Alexander* the Great (d. 323 BC), a great medical school was founded in Egypt at the court of King Ptolemy, at his capital, *Alexandria*, at the mouth of the Nile. The King's main cultural creations were the Alexandrian Library and the Museum (Sanctuary of the Muses), which installed Greek learning in a new Egyptian environment—Archimedes, Euclid, and the astronomer Ptolemy were soon to teach there. The Library became a wonder of the scholarly world, eventually containing, it was said, 700,000 manuscripts, and other facilities including an observatory, zoological gardens, lecture halls and rooms for research.

The two earliest teachers at the Alexandrian medical school were also its greatest—*Herophilus* of Chalcedon (ca. 330-260 BC) and his contemporary, *Erasistratus* of Chios (ca. 330-255 BC). Their writings having been lost, we know about them only through later physicians.

Herophilus was the first to dissect cadavers in public. He was a student of *Praxagoras* of Cos (fl. 340 BC), who had improved *Aristotelian* anatomy by distinguishing arteries from veins, but who saw the arteries as air tubes, similar to the trachea and bronchi, a common error because arteries are devoid of blood in corpses. Herophilus observed that the coats of arteries were much thicker than those of the veins, thus he speculated that the arteries were filled not with air but with blood. *Herophilus* wrote at least eleven treatises, discovering and naming the prostate and the duodenum (from the Greek for twelve fingers, the length of gut he found). He also wrote on the pulse as a diagnostic guide and on therapeutics, ophthalmology, dietetics, and midwifery. He recognized the brain as the central organ of the nervous system and the seat of intelligence, extending the knowledge of the parts of the brain, certain of which still bear titles translated from those given by him. He was the first to grasp the nature of the nerves, which he distinguished as motor and sensory, though he did not separate them clearly from tendons.

Erasistratus surmised that every organ is formed of a threefold system of "vessels" —veins, arteries, and nerves, dividing indefinitely. These, plaited together, were postulated to make up the tissues. In the brain, *Erasistratus* observed convolutions, noting that they were more elaborate in man than in animals and associating this with higher intelligence. He distinguished between cerebrum and cerebellum, and is often regarded as an early mechanist because of his model of bodily processes—for instance, digestion involved the stomach grinding food. He was opposed to intrusive remedies such as venesection, and his favourite therapeutic measures were regulated exercise, diet, and the vapour bath—very much in the Hippocratic tradition.

Ancient Roman medicine

Roman tradition held that one was better off without doctors. According to Cato (234-149 BC), citizens had no need of professional physicians because Romans were hale and hearty, unlike the effete Greeks. Apparently Romans enjoyed bad-mouthing Greek physicians. Thus the Romans despised medicine as a profession but this did not prevent them from making use of Greek physicians or even of the services of their own slaves. *Galen of Pergamum* (130-200 AD) tells us that in his time, large cities such as *Rome* and *Alexandria* swarmed with specialists who also travelled about from place to place. *Martial* (40-104 AD) mentions some of them in an epigram: "Cascellius extracts and repairs bad teeth; you, Hyginus, cauterize ingrowing eyelashes; Fannius cures a relaxed uvula without cutting; Eros removes brand marks from slaves; Hermes is a very Podalirius for ruptures." Under the Empire, military medicine was highly organized—every cohort had its surgeon, and surgeons of a higher grade were attached to the legions as consultants. Army surgeons ranked as noncombatants and enjoyed many privileges.

In the Roman empire, the earliest scientific teacher was a Greek, Asclepiades of Bithynia (124-40 BC). *Asclepiades* ridiculed the Hippocratic expectant attitude as a mere "meditation on death," and urged active measures that the cure might be "seemly, swift and sure." Though outside the mainstream, his medical practice is interesting as a modification of the atomic or corpuscular theory, according to which disease results from an irregular or inharmonious motion of the corpuscles of the body. His pupils were numerous, constituting the Methodical school, but his available therapeutic tools were few—he trusted mainly to changes of diet, friction, bathing, exercise, and occasionally emetics, bleeding, and wine. He was also the first to use music in the treatment of the insane.

Back in the mainstream, *Galen of Pergamum* (130-200 AD) provided the final medical synthesis of antiquity and the effective medical standard for the next 13 centuries. His first medical appointment was as surgeon to the Roman gladiators. He later traveled to Rome and wrote extensively on anatomy, physiology and practical medicine. His fame is due in part to his prolific pen some 350 authentic titles ranging in topic from the soul to bloodletting polemics survive, about as much as all Greek medical writings together. Galen was a flamboyant character. One of his party tricks, revealing his genius for self-advertisement as well as experiment, was to sever the nerves in the neck of a pig. As these were severed, one by one, the pig continued to squeal; but when Galen cut one of the laryngeal nerves the squealing stopped, impressing the crowd.

Galen justified venesection in terms of his elaborate pulse lore. Written in the early 170s, his sixteen books on the pulse were divided into four treatises, each four books long. In one of these treatises, he explains how to take the pulse and to interpret it, raising key questions, for example: How was it possible to tell whether a pulse was full, rapid, or rhythmical? Such questions he resolved partly from experience and partly by reference to earlier authorities. Galen developed a characteristic physiological scheme that remained in vogue until the 17th century. It supposes three types of so-called spirits associated with three types of the activity of living things. These were the natural spirits formed in the liver and distributed by the veins, the vital spirits formed in the heart and distributed by the arteries, and the animal spirits formed in the brain and distributed by the nerves. Galen's system was an admirable if factually flawed working hypothesis, based on much experimental evidence, and he presented his work as "perfecting" the legacy of Hippocrates.

As in Greece, medicine remained personal in Rome. No medical degrees were conferred or qualifications required. In the absence of colleges and universities, the private face-to-face nature of medical instruction encouraged fluidity and diversity, Students attached themselves to an individual teacher, sitting at his feet and accompanying him on his rounds. Many different sorts of medical care were available, and self-help was universal. Celsus' *On Medicine* was written for a non-professional readership as willing to wield the scalpel as the plough. Disease explanations changed little. Public authorities still ascribed famines and pestilences to the gods, and Galen was silent on contagion. The essential trilogy of classical medicine remained dietetics, exercise, and drugs, accompanied by light surgery.

Medicine in the Middle ages

The passage from the glorious days of Rome to the Middle Ages was often violent, especially in the West, with wave after wave of barbarian onslaughts from the East. These culminated in the sack of the Eternal City by Alaric's Goths in 410 AD, which effectively ended the western empire and frayed the thread of learned medicine. Thus Galen had no effective successor. Indeed, medieval medicine may be summed up as a corrupted version of Galenism. The true scientific tradition did not reappear in the West until the 16th century, after a lengthy incubation in the Islamic world. For example, in England the Venerable Bede (ca. 672-735) and his monks possessed many medical writings, and knowledge of plant remedies was extensive, but the English healer used chants and charms, predicated on the belief that certain diseases and bad luck were caused by darts shot by elves, while other ailments involved a "great worm," a term applied to snakes, insects, and dragons.

By contrast with the naturalist focus of *Hippocratic* and *Galenic medicine*, healing became more authoritarian and intertwined with religion, for the rising Church taught that there was a supernatural plan and purpose to everything, including sickness and death. *Christian* and *Jewish* healing traditions became more prominent. Disease could be cured by prayers or by invoking the names of saints, by exorcism, by amulets or number magic, or by transferring the sickness to animals, plants, or to the soil. Certain maladies such as leprosy were associated with the Almighty's punishments for sin, according to the Book of Leviticus:

"When a man shall have in the skin of his flesh a rising [a swelling], a scab, or bright spot, and it be in the skin of his flesh like the plague [the spots] of leprosy...and the plague in sight be deeper than the skin of his flesh it is a plague of leprosy: and the priest shall look on him, and pronounce him unclean."

Such polluting diseases were curable by the Lord alone, encouraging some Jewish people to reject human medicine in favour of the divine, citing the fate of *King Asa* (ca. 914-874 BC), who "sought not the Lord, but his physicians," and whose foot sores consequently worsened, and he died. On the other hand, while *Jewish* dietary rituals (e.g. kosher food) are principally expressions of religious precepts about pollution and purification, a useful practical effect is to limit exposure to foodborne diseases such as trichinosis.

Many of the distinguished surgeons of the Middle Ages were clerics, but the practice of medicine and surgery by members of the Church was not favoured by the hierarchy. Many laws were passed against the practice of medicine for worldly profit. In 1215 the Fourth Lateran Council forbade all sub-deacons, deacons, or priests to practice that part of surgery that had to do with burning and cutting. Finally,

Pope *Honorius III* (d. 1216) prohibited all persons in holy orders from practicing medicine in any form, which meant that the educated classes were prohibited from performing any type of surgery.

In Europe in the Middle Ages, practitioners unsuccessfully sought to cure wounds by treating only the weapons that caused them, using an ointment on the offending knife or sword known as a weapon salve. Physicians practiced urine-gazing (uroscopy), but proper chemical analysis of urine did not develop until the 18th century. Medieval medicine is the source of many humorous "remedies" in contemporary diatribes against modern medicine—such as being strapped into a halter and walked around a pigsty three times for mumps, drinking water from the skull of a bishop to treat rheumatism, or nuzzling a mouse to cure the common cold. But we must recognize that such treatments were not put forth as reasonable science. Rather, they serve as a reminder of the results of injecting mystical or religious elements into medicine as a substitute for good science.

The Middle Ages are also remarkable for their plagues, which became more numerous as people crowded closer and closer together in urban centers of increasing population density. *Plague* records from China during this period often show 50%-70% mortality rates, and China probably served as the original source of the greatest of the bubonic plagues that swept through Europe in the 14th century, known as the Black Death. The disease broke out in 1346 among the armies of a Mongol prince who laid siege to the trading city of Caffa in the Crimea. This compelled his withdrawal, but not before the disease had entered Caffa, whence it rapidly spread by ship throughout the Mediterranean.

The initial shock in 1346-1350 was severe. Die-offs varied widely, with some small communities experiencing total extinction and others, such as Milan, being spared entirely. The lethal effect of the plague may have been enhanced by the fact that it was propagated not solely by the bites of fleas carried on infected rodents, but also was transmitted person to person as a result of inhaling bacillus-filled droplets that had been coughed or sneezed into the air by an infected individual. Lung infections of this kind were observed to be 100% lethal in Manchuria in 1921, the only time modern medicine has directly observed airborne plague communication, so it is tempting to assume a similar mortality for pneumonic plague in 14th century Europe. Mortality rates for sufferers from bubonic infection transmitted by flea bite varies from 30% to 90%. All told, the best estimate of

European plague-induced mortality is that about one-third of the population died during the initial five-year period.

The plague returned in the 1360s, the 1370s, and thereafter, and European population declined irregularly as a result, reaching a low point in England sometime between 1440-1480. People learned to minimize infection risk via quarantine, a practice which stemmed from Biblical passages prescribing the ostracism of lepers. Plague sufferers were treated as though they were temporary lepers, with a standard 40-day quarantine for people and incoming ships at all major ports. However, since the role of fleas and rats in disease propagation remained unknown until the end of the 19th century, quarantine measures were often ineffectual. Through the 17th century, occasional plague outbreaks that carried off up to a third or a half of a city's population in a single year were considered normal. For example, Venetian statistics show that in 1575-77 and again in 1630-31, a third or more of the city's population died of plague.

Pre-modern medicine was powerless against bubonic plague. Before antibiotics reduced the disease to triviality in 1943, the average mortality rate was 60%-70% of those affected, despite all that the best hospital care could accomplish. The last recorded outbreak of plague that ran its course without benefit of penicillin and related antibiotics (which destroy the infection rapidly) occurred in Burma in 1947, with 78% lethality.

Renaissance and Pre-modern medicine

While internal medicine languished through the Middle Ages, the surgeons of the Renaissance gained wide experience during the many religious wars of the period. They had many new problems to face, including the treatment of wounds caused by firearms. The French had employed gunpowder at the siege of Puiguillaume in 1338, and cannon were used by the English at the Battle of Crecy in 1346. Gunshot wounds at that time were caused by large missiles of low velocity that caused ragged wounds and carried pieces of clothing into the tissues. These wounds were severe and very liable to become septic. The universal belief among contemporary surgeons was that gunpowder itself was venomous. To neutralize the effect of this venom, the general practice was to cauterize the wound by injecting boiling oil.

The first man to break away from this old doctrine was *Ambroise Pare* (1510-90). Pare came to Paris in 1532 as an apprentice to a barber-surgeon and then moved to the great Hotel Dieu as resident surgeon. In that immense medieval hospital, the only one in Paris at the time, he gained great experience, and in 1536 he began his career as a military surgeon. As described in his book *The Apologie and Treatise*, he recounts how, during his first campaign as a greenhorn military surgeon in Turin in 1537, he had run out of boiling oil, the established treatment for gunpowder wounds, just after French troops had captured the castle of Villaine. So in place of boiling oil, he applied:

"...a digestive of yolk of eggs, oil of roses, and turpentine. In the night I could not sleep in quiet, fearing some default in not cauterizing, that I should find those to whom I had not used the burning oil dead impoisoned; which made me rise very early to visit them, where beyond my expectation I found those to whom I had applied my digestive medicine, to feel little pain, and their wounds without inflammation or tumour, having rested reasonable well in the night; the other to whom was used the said burning oil, I found them feverish with great pain and tumour about the edges of their wounds. And then I resolved with myself never so cruelly to burn poor men wounded with gunshot."

Pare also went on to show that bleeding after amputations should be arrested, not by the terrible method of the indiscriminate use of the red-hot cautery, but by simple tying of the blood vessels. His most famous phrase, so reminiscent of the old Hippocratic school of thought, was: "I dressed the wound, and God healed him."

In 1633 appeared the earliest book on first-aid for the injured, by one *Stephen Bradwell*, although the proffered advice sounds impractical and a bit odd to modern ears. For example, the treatment for the "Biting of a Madde Dogge" is to throw the patient into water. "In doing this, if he cannot swim, after he hath swallowed a good quantity of water, take him out again. But if he be skilful in swimming, hold him under the water a little while till he have taken in some pretty quantity." This procedure may not be wholly irrational—standard 20th century first aid for dog bites includes a thorough cleansing of the wound with water.

Venesection remained a popular 17th century universal remedy. As described by the surgeon *Richard Wiseman* (1622-1676), "a gentleman of about thirty years of age coming out of *Hertfordshire* through Tottenham and riding upon the causeway near an inn, one emptying a chamber pot out of the window as he was passing by, his horse started and rushed violently between a signpost and a tree which supported part of the sign. The poor gentleman was beaten off his horse and lay stunned upon the ground." A barber-surgeon was hastily summoned but nothing much was done for the injured man until Wiseman arrived, whereupon:

"I found the gentleman lying upon the ground, the people and chirurgeon gazing upon him. I felt his pulse much oppressed, the right brow bruised and inquired whether they had bled him blood. The chirurgeon replied that he had opened a vein in his arm but it would not bleed. I replied, we must make him bleed through it by splitting his veins. Turning his head on one side, I saw the jugular vein on the bruised side turgid and opened it. He bled freely. After I had taken about twelve ounces, the blood ran down from his arm which had been opened before and would not bleed. We bled him till he came to life, and then he raved and struggled with us."

The patient's injuries were dressed and he was subjected to further bleedings, but evidently made a good recovery. Even by the 18th century, the traditional surgeon's day-to-day business eschewed high-risk operations like amputations; rather, it was a round of minor procedures such as venesection, lancing boils, dressing skin abrasions, pulling teeth, managing whitlows, trussing ruptures, and treating skin ulcers. The fatality rates of these procedures were low, for surgeons understood their limits, and the repertoire of operations they attempted was small, because of the well-known risks of trauma, blood loss, and sepsis. Internal disorders were treated not by the knife but by medicines and management, since major internal surgery was unthinkable before anesthetics and antiseptic procedures. Improvements did occur in certain operations such as lithotomy. *William Cheselden* (1688-1752), a great British surgeon of the 18th century, perfected a technique which enabled him to remove a stone in the bladder in one minute (his record time was 54 seconds), thus reducing mortality from about 50% to under 10%. Cheselden's results were not bettered until almost the end of the 19th century.

In the 17th century, internal medical treatment was frequently overdone on those affluent enough to afford it. Critics often denounced physicians as meddlesome, capriciously practicing an often dangerous polypharmacy—a blunderbuss approach. The deathbed of *Charles II* (1630-1685) of England was a conspicuous case of such medical overkill; after the king had suffered a stroke, his doctors moved in, and *Sir Raymond Crawfurd* (1865-1938) recreated the scene:

"Sixteen ounces of blood were removed from a vein in his right arm with immediate good effect. As was the approved practice at this time, the King was allowed to remain in the chair in which the convulsions seized him. His teeth were held forcibly open to prevent him biting his tongue. The regimen was, as Roger North pithily describes it, first to get him to wake, and then to keep him from sleeping. Urgent messages had been dispatched to the King's numerous personal physicians, who quickly came flocking to his assistance; they were summoned regardless of distinctions of creed and politics, and they came. They ordered cupping-glasses to be applied to his shoulders forthwith, and deep scarification to be carried out, by which they succeeded in removing another eight ounces of blood. A strong antimonial emetic was administered, but as the King could be got to swallow only a small portion of it, they determined to render assistance doubly sure by a full dose of Sulphate of Zinc. Strong purgatives were given, and supplemented by a succession of clysters. The hair was shorn close, and pungent blistering agents were applied all over his head. And as though this were not enough, the red-hot cautery was requisitioned as well."

One of the dozen attending physicians noted with pride that "nothing was left untried"; the King graciously apologized for being "an unconscionable time a-dying."

Meanwhile, the common citizen experienced poor health exacerbated by the many new dangers attending the Industrial Revolution. In 1775, *Percivall Pott* (1714-1788) pointed out that boy chimneysweeps developed scrotal cancer, due to soot irritation. In his *Condition of the Working Classes in England* (1844), *Friedrich Engels* (1820-1895), a Manchester factory owner as well as *Karl Marx's* collaborator, described workers who were "pale, lank, narrow-chested, hollow-eyed ghosts," cooped up in houses that were mere "kennels to sleep and die in." In 1832, the Leeds physician *Charles Turner Thackrah* (1795-1833) published *The Effects of Arts, Trades, and Professions on Health and Longevity*, documenting the diseases and disabilities of various occupations. Apart from factory workers, among those most exposed to harmful substances were cornmillers, maltsters, coffee-roasters, snuff-makers, rag-pickers, papermakers and feather-dressers. Tailors were so subject to anal fistulas that they set up their own "*fistula clubs*." Thackrah's overall verdict was bleak: "Not 10% of the inhabitants of large towns enjoy full health."

The single worst malady cultivated in populous cities was tuberculosis (TB), a disease characterized by fever, night sweats, and hemoptysis (coughing up blood), called "consumption" because victims were almost literally consumed. By 1800 TB was proclaimed the most common disease, and in 1815 *Thomas Young* (1773-1829) surmised that tuberculosis brought a premature death to one in four in the general population. Autopsies conducted in the chief Paris hospitals recorded TB as the cause of death in some 40% of cases. In the continental U.S. as late as 1890, the corresponding percentage was about 13%, and tuberculosis was still the leading cause of death, though the data are partially suspect because cases of lung cancer were sometimes reported as "consumption".

One important 18th century improvement in internal medicine which decisively saved many lives was the introduction of inoculation and vaccination against smallpox. Smallpox, "the speckled monster," had become virulent throughout Europe and in bad years accounted for about 10% of all deaths; *Queen Mary of England* (1662-1694), *Louis XV of France* (1710-1774), and Queen Anne's son and sole surviving direct heir (d. 1700) died of it. Doctors had long been aware of the immunizing properties of an attack, and smallpox inoculation seems to have been known and practiced for centuries at a folk level throughout Arabia, North Africa, Persia, and India. Reports of a more elaborate Chinese method, involving the insertion of a suitable infected swab of cotton inside the patient's nostril, reached London in 1700. But it was a report from *Mary Wortley Montagu* (1689-1762), wife of the British consul in Constantinople, that Turkish women held smallpox parties at which they routinely performed inoculations with the aim to induce a mild dose so as to confer lifelong protection without pockmarking, that hastened acceptance in the rural medical community. The usual method was to transfer the infection by introducing matter from a smallpox pustule into a slight wound made in the patient's skin. Occasionally the patient developed a severe case of smallpox from such treatment, and some died. But usually the symptoms were slight—a few score of pox only—and immunity proved equivalent to that resulting from contracting the disease naturally.

Edward Jenner (1749-1823), an English country doctor who performed such inoculations, noticed that cowpox, a cattle disease occasionally contracted by humans, particularly dairy maids, also conferred immunity against smallpox. Suspecting that it might be possible to produce this immunity by arm-to-arm inoculation from the cowpox pustule, and surmising it would be safer than inoculation from smallpox pustules directly, since in humans cowpox was benign, Jenner tried the experiment, and it worked. In 1798 he published his discovery in *An Inquiry into the Causes and Effects of the Variolae Vaccinae.*

By 1799 over 5000 individuals had been vaccinated in England and abroad the practice was taken up remarkably swiftly, being made compulsory in Sweden and supported by *Napoleon*, who had his army vaccinated. For the first time in history, organized medicine began to contribute to human population growth in a statistically significant fashion.

The Napoleonic Wars spurred new attempts to treat battle-wounded soldiers in a more timely manner. Traditionally, the wounded were left on the field unattended until the end of battle, but Napoleon's chief surgeon *Dominique Jean Larrey* (1766-1842) introduced the use of rudimentary carts called ambulances volantes (little more than horsedrawn rickshaws) as the first "ambulances" to evacuate and transport wounded soldiers from the field to nearby aid stations, even while the battle raged on. In 1792, Larrey organized the first air evacuation, by hot air balloon.

The use of ambulances didn't catch on until the late 1800s; until then, anyone injured in the streets of Paris, London, New York or Boston depended on the kindness of strangers or a nearby business shop for a place to rest until a doctor could be summoned. The first "modern" ambulance appeared in the city of Cincinnati in 1865, but the first true city ambulance system was developed in association with Bellevue Hospital in New York City in 1866, receiving 1500 requests for transport in its first three years of service. These horse-drawn ambulances, usually provided by local mortuaries, carried a driver and a surgeon, who was on board mainly to pronounce a patient's death at the scene or upon arrival at the hospital, since little could be done for the seriously injured. The surgeon kept meticulous notes on the ride, recording the time of the call, transport and arrival times, and any other details that "a coroner's jury might possibly require".

The period also saw the development of many simple diagnostic tools that are taken for granted today. For example, a French physician, *Rene Theophile Hyacinthe Laennec* (1781-1826), in his *Treatise On Mediate Auscultation* (1819), described pathological lesions found in the chest at autopsy and showed how they correlated with disease detected in living patients, establishing for the first time the modern concept of clinicopathological correlation, the cornerstone of modern diagnosis. Laennec also developed an instrument that he named a "stethoscope" to assist him in his examination of patients, especially female patients, against whose chests the direct placing of a male ear was socially taboo. The original device was a straight wooden tube;

by mid-century rubber tubing was introduced to create a flexible monaural stethoscope, and in 1852 an American physician, *George P. Cammann* (1804-1863), devised our familiar two-ear instrument.

The clinical thermometer is of like vintage. *Galileo* (1564-1642) invented the first thermometer in the late 16th century, but it was not applied to medicine. Early medical thermometers in the 18th century were a foot long and difficult to use at the bedside, and were reportedly carried under the arm "as one might carry a gun." The short clinical thermometer was devised by *Sir Clifford Allbutt* (1836-1925) in the 1860s, and was widely used during the American Civil War (1861-1865). The classical work in temperature diagnostics was *Carl Wunderlich's* (1815-1877) *The Temperature in Diseases*, published in 1868, which presented data on nearly 25,000 patients and analyzed temperature variations in 32 diseases, showing that temperature readings could differentiate fevers. Other devices emerged later to measure pulse and blood pressure. In 1854, *Karl Vierordt* (1818-1884) created the sphygmograph, a pulse recorder usable for routine monitoring on humans. Blood pressure was measured using the familiar inflatable band wrapped around the upper arm, called the sphygmomanometer, whose basic design was established in 1896 by *Scipione Riva-Rocci* (1863-1937). The hypodermic syringe was invented in 1853.

Biochemistry also began to play an increasing diagnostic function. In the 18th century, *Matthew Dobson* (1784) developed tests for diabetes. In 1827, *Richard Bright* (1789-1858) showed show the kidney complaint subsequently called Bright's disease could be diagnosed by a single, simple chemical test. Chemical analysis was crucial to Alfred Becquerel's (1814-1862) urinalysis studies in 1841, establishing the average amounts of water, urea, uric acid, lactic acid, albumin, and inorganic salts secreted over 24 hours, and correlating these with various disease conditions. In 1859, *Alfred Garrod* (1819-1907) devised a simple chemical test pathognomonic for gout.

Fully invasive surgery

The fully invasive surgery of the 19th and 20th centuries rests upon the triple foundation of anatomy, anesthesia, and asepsis. Each has an interesting story, described below, presaging the emergence of the rational-scientific approach to medical treatment.

Anatomy

In surgical practice, lack of accurate anatomical knowledge had long been a great obstacle. Dissection of the human body was first practiced systematically at the great medical school of Alexandria, which flourished from about 300 BC until the death of the last ruler of Ptolemaic Egypt, Cleopatra, in 30 BC. After the decline of Alexandria, dissection was carried on at a few other centers in the Middle East, but in the first two centuries of the Christian era human bodies were replaced on the dissection table by those of apes and other animals. The anatomical knowledge gained by Galen and others from the dissection of animals was an adequate guide for the simple operative procedures carried out at this time because the abdomen, the chest, and the head were rarely opened by the surgeon's knife.

Anatomical demonstrations of a kind were introduced into some Italian medical schools early in the 14th century but their main purpose was to serve as an aid in memorizing what Galen had written a thousand years before. During the late Middle Ages, and for a long time after, the procedure was for the professor to read from some second- or third-hand manuscript version of Galen while a demonstrator pointed to the part under discussion with a wand. As the text of Galen was often based on the dissection of an ape or pig, there were naturally many occasions when the anatomical structure under examination did not correspond with Galen's description.

Some of the great artists took up the scientific study of anatomy, but the true founder of modern anatomy was *Andreas Vesalius* (1514-1564), a native of Brussels who studied medicine in Paris and later taught surgery and anatomy at Padua and Bologna. Vesalius was filled with a passionate desire for anatomical study, and many stories are told of the great risks which he took in obtaining material, including, it is reported, graverobbing. On one occasion he stole the skeleton of a criminal which was hanging on a gallows outside the city wall of Louvain and the trophy proved of great value in his studies. His public dissections during his seven years in Padua drew enormous crowds of students. In 1543 he published his 355-page great work, *De Humani Corporis Fabrica Libri Septem (On the Fabric of the Human Body)*, the outstanding and precise illustrations in which were copied and plagiarized by scholars for more than 100 years, and could be used for teaching even today.

For the first time in the history of medicine, doctors had at their disposal a detailed and accurate anatomical text with illustrations from the hand of a great artist. This book is the foundation stone of anatomy—indeed of all modern medicine—because without a sound knowledge of the structure of the body there can be no real understanding of the body's functions in health and disease.

A tremendous shortage of dead human bodies for dissection, to teach anatomy, remained a problem for centuries. In Edinburgh in 1827, an old man died in *William Hare's* (1792-1870) boardinghouse; assisted by his lodger William Burke (1792-1829), the two men bypassed the grave and sold the body directly to anatomists. Spurred by success, they turned to murder, luring victims and suffocating them to avoid signs of violence. Sixteen were done to death and their bodies sold, fetching 7 pounds apiece, before Burke and Hare were brought to justice in 1829. Hare turned King's evidence and Burke was hanged. The last cadaver was found in the dissecting room of a respected anatomist, *Robert Knox* (1791-1862). Despite his cries of innocence, an incensed crowd burned down his house and he fled to London, his career in ruins, and eventually *Knox* died in obscurity.

To help pass the English Anatomy Act of 1832 (which awarded to doctors the "unclaimed bodies" of paupers) and to dispel public concern about dissection, the body of the great English philosopher and jurist, *Jeremy Bentham* (1748-1832), in accordance with his directions, was dissected in the presence of his friends. The skeleton was then reconstructed, supplied with a wax head to replace the original (which had been mummified), dressed in Bentham's own clothes and set upright in a glass-fronted case. Both this effigy and the head are preserved at University College, London, to this day.

Anesthesia

Pain did not prevent surgery but made it almost unbearable, and the accompanying trauma often proved dangerous. Before the anesthetic era, which commenced in the late 1840s, surgical operations were agonizing. Of course, if the patient had a broken leg or amajor wound, there was no choice but submit to a surgeon's knife. But non-emergency elective operations would only be undergone if the condition itself was so painful or life-threatening that the victim could even consider allowing surgery. In this event, patients would choose a surgeon with the best reputation for quickness—a limb might be removed or a bladder stone evacuated in a couple of minutes. Progress in technique was often rapid. For example, in 1824, *Astley Cooper* (1768-1841) took 20 minutes to amputate a leg through the hip joint; ten years later, *James Syme* (1799-1870) was doing it in just 90 seconds.

In pre-anesthetic days, operations were rushed through at lightning speed and under conditions of appalling difficulty. The most hardened surgeons had to steel themselves to perform operations which they knew would cause agony to their patients and nerve-wracking distress to themselves. It is hard for any 20th century inhabitant of an industrialized nation to imagine what a major surgical operation must have meant to the patient in the days before anesthesia. The following is a personal account by a male patient who suffered the removal of a stone from the bladder by *Henry Cline* (1750-1827), surgeon to St. Thomas's Hospital and one of the leading operators of the day, on 30 December 1811, just three decades before the widespread adoption of anesthesia:

"My habit and constitution being good it required little preparation of body, and my mind was made up. When all parties had arrived I retired to my room for a minute, bent my knee in silent adoration and submission, and returning to the surgeons conducted them to the apartment in which the preparations had been made. The bandages &c. having been adjusted I was prepared to receive a shock of pain of extreme violence and so much had I overrated it, that the first incision did not even make me wince although I had declared that it was not my intention to restrain such impulse, convinced that such effort of restraint could only lead to additional exhaustion. At subsequent moments, therefore I did cry out under the pain, but was allowed to have gone through theoperation with great firmness."

"The forcing up of the staff prior to the introduction of the gorget gave me the first real pain, but this instantly subsided after the incision of the bladder was made, the rush of urine appeared to relieve it and soothe the wound."

"When the forceps was introduced the pain was again very considerable and every movement of the instrument in endeavoring to find the stone increased. Still, however, my mind was firm and confident, and, although anxious, I was yet alive to what was going on. After several ineffectual attempts to grasp the stone I heard the operator say in the lowest whisper, "It is a little awkward, it lies under my hand. Give me the curved forceps," upon which he withdrew the others. Here, I think, I asked if there was anything wrong—or something to that purport — and was reanimated by the reply conveyed in the kindest manner, "Be patient, Sir, it will soon be over." When the other forceps was introduced I had again to undergo the searching for the stone and heard Mr. Cline say, "I have got it." I had probably by this time conceived that the worst was over; but when the necessary force was applied to withdraw the stone the sensation was such as I cannot find words to describe. In addition to the positive pain there was something peculiar in the feel. The bladder embraced the stone as firmly as the stone was itself grasped by the forceps; it seemed as if the whole organ was about to be torn out. The duration, however, of this really trying part of the operation was short and when the words "Now, Sir, it is all over" struck my ear, the ejaculation of "Thank God! Thank God!" was uttered with a fervency and fulness of heart which can only be conceived....I never heard what was the precise duration of the operation but conceive it to have been between twelve and fifteen minutes."

And now, a woman's point of view. In 1810, Napoleon's famed military doctor Dominique *Jean Larrey* performed a radical mastectomy without anesthetic on the popular female novelist *Fanny Burney* (1752-1840). Burney later wrote a long account of the operation which, despite the excruciating agony, she believed had nevertheless saved her life:

"M. Dubois placed me upon the Mattress, & spread a cambric handkerchief upon my face. It was transparent, however, & I saw through it that the Bed stead was instantly surrounded by the 7 men and my nurse. I refused to be held; but when, bright through the cambric, I saw the glitter of polished steel — I closed my eyes..."

"Yet — when the dreadful steel was plunged into the breast — cutting through veins—arteries—flesh—nerves—I needed no injunctions not to restrain my cries. I began a scream that lasted unintermittingly during the whole time of the incision—& I almost marvel that it rings not in my Ears still! so excruciating was the agony."

"When the wound was made, & the instrument was withdrawn, the pain seemed undiminished, for the air that suddenly rushed into those delicate parts felt like a mass of minute but sharp & forked poniards [small pointed daggers], that were tearing at the edges of the wound. But when again I felt the instrument, describing a curve, cutting against the grain, if I may so say, while the flesh resisted in a manner so forcible as to oppose & tire the hand of the operator, who was forced to change from the right to the left —then, indeed, I thought I must have expired, I attempted no more to open my eyes....The instrument the second time withdrawn, I concluded the operation over—Oh no! presently the terrible cutting was renewed—& worse than ever, to separate the bottom, the foundation of the dreadful gland from the parts to which it adhered...yet again all was not over...."

When did the practice of anesthesia begin? The Herbal of Dioscorides (ca. 40-90 AD) contains specific directions for giving a decoction (boiled extraction) of mandragora "to such as shall be cut or cauterized," one of the earliest references to surgical anesthesia.

Bernard de Gordon (ca. 1260-1308) tells us that the Salernitans rubbed up poppy seed and henbane and used them as a plaster to deaden the sensibility of a part to be cauterized. *Arnold of Villanova* (1235-1311) gives the following recipe:

"To produce sleep so profound that the patient may be cut and will feel nothing, as though he were dead, take of opium, mandragora bark, and henbane root equal parts, pound them together and mix with water. When you want to sew or cut a man, dip a rag in this and put it to his forehead and nostrils. He will soon sleep so deeply that you may do what you will. To wake him up, dip the rag in strong vinegar."

Some of the surgical textbooks of the Middle Ages contain references to anesthetic sponges which were prepared by soaking them in various herbs reputed to have soporific properties. The favourite herb for this purpose was the mandrake. Another simple method of producing analgesia used intermittently from early times was compression. Writing in 1564 about the various uses of the tourniquet, the French surgeon *Ambrose Pare* noted that "it much dulls the sense of the part by stupefying it." Amusingly, reliable witnesses claim that as late as the 19th century, a method of anesthesia practiced at the Imperial Court of China was to "knock the patient out by a sudden blow on the jaw."

The almost complete absence of any mention of pain-relieving drugs in medical literature of the post-medieval period is not easy to explain. It is, however, probable that the action of crude concoctions employed in early times was very uncertain, and that drugged sleep often ended in death. The active ingredients of the many herbs used in medicine had not been isolated and it would have been very difficult to regulate dosages reliably.

The story of inhalation anesthesia begins in 1799 when *Sir Humphry Davy* (1778-1829) recorded the effects produced by the inhalation of nitrous oxide. He breathed various concentrations of the gas and noted that a headache and the pain associated with the cutting of a wisdom tooth were relieved. Demonstrations of the effects of nitrous oxide were frequently given, bladders filled with "laughing gas" being passed around at lectures. Gas inhalation became a popular party game. A little book of 1839 contains adescription of the "irresistibly ridiculous" sight of a large room filled with persons each of whom was sucking from a bladder. As the gas began to take effect, "some jumped over the tables and chairs; some were bent on making speeches; some were very much inclined to fight; and one young gentleman persisted in attempting to kiss the ladies." At about the same time, "ether frolics" became equally popular.

In January 1842, *William E. Clarke* (b. 1818), a young American physician of Rochester, New York, who had acquired some knowledge of ether by attendance at ether frolics, administered the chemical on a towel to a Miss Hobbie who then had one of her teeth extracted painlessly. So far as is known this was the first use of ether for a dental or surgical operation. In March 1842, *Crawford W. Long* (1815-1878) of Danielville, Georgia, who had also witnessed ether frolics "enjoying sweet kisses from the girls," successfully removed a small tumour from the neck of a patient under the influence of ether. *Horace Wells* (1815-1848), a dentist of Hartford,Connecticut, attended a public demonstration of the effects of nitrous oxide in December 1844, and the day after he administered the gas to himself and had one of his own teeth pulled out by a colleague. Afterwards, Wells wrote: "I didn't feel it so much as the prick of a pin." His former partner, *William Thomas Green Morton* (1819-1868), also introduced ether into his dental practice in 1846. By February 1847, the Lancet and other medical journals were reporting anesthetic operations from all parts of Great Britain, and ether had been used in most European countries. In June 1847 the news reached South Africa and a leg was amputated painlessly by *W.G. Atherstone* of Grahamstown.

Sir James Young Simpson (1811-1870), professor of surgery at Edinburgh, introduced chloroform in 1847. Tradition has it that Simpson had been testing chemicals with his assistants when somebody upset a bottle of chloroform; upon bringing in dinner, Simpson's wife found them all asleep. Unlike ether, chloroform did not irritate the lungs or cause vomiting, and was powerful and easy to administer. In April 1853, *Queen Victoria* (1819-1901) took chloroform for the birth of Prince Leopold; *John Snow* (1813-1858) administered the anesthetic. Protests followed—some objections were religious (e.g., the Bible taught that women were supposed to bring forth in "travail and pain") but most were medical, putatively on grounds of safety but ringing with naturophilia: "In no case could it be justifiable to administer chloroform in perfectly ordinary labour," complained the Lancet. Incredibly, some early 19th century surgeons believed that using anesthesia for an operation would weaken a patient's character.

Of course, general anesthesia could indeed prove dangerous, and deep unconsciousness was unnecessary for less invasive procedures, so the search was on for substances that would numb a particular area for local surgery. Cocaine was isolated in 1859 and was first used in ophthalmologic procedures by Carl Koller (1857-1944). Cocaine became the first widely-used local anesthetic, synthesized in 1885 by the Merck drug company.

Germ theory and antisepsis

By the middle of the 19th century, pain had been banished from surgical operations, but one grave danger still faced every patient submitting himself to the surgeon's knife. This was the ever-present risk of sepsis (infection). Hospital diseases such as erysipelas, pyemia, septicemia and gangrene, were rife. In the 1850s the death rate after amputations varied from 25%-60% in different countries and in military practice it reached the appalling figure of 75%-90%. The first ovariotomies, which were the first abdominal operations performed on a fairly large scale, had a mortality rate of more than 30% even in the most expert hands. That all these diseases were due to some form of "contagion" had long been suspected, but the general view was that whatever agent was responsible was generated spontaneously in wounds. Alternatively, it was theorized that air itself was responsible for suppuration and many attempts were made to exclude the air from wounds by means of elaborate dressings.

Some medical men had postulated the existence of minute particles in the air which carried contagion, the so-called "germ theory." In 1546, *Girolamo Fracastoro* of Verona (1478-1553) proposed "seminaria, the seeds of disease which multiply rapidly and propagate their life," minute bodies passing unseen from the infector to the infected by contact, by clothing or utensils, and by infection at a distance through the air.

What made the germ theory of contagion so difficult to accept was that no one could see the supposed microbes. Magnifying lenses were used in ancient times, and by the beginning of the 17th century they had been combined in a tube to make the compound microscope. The first man to employ the microscope in investigating the causes of diseases was probably *Athanasius Kircher* (1601-1680), a learned Jesuit priest. In 1658 Kircher described experiments upon the nature of putrefaction, showing how maggots and other living creatures developed in decaying matter. He also claimed to have found in the blood of plague-stricken patients "countless masses of small worms, invisible to the naked eye." It is impossible that he could have seen the plague bacillus with the very low power microscopes at his disposal, but he may have seen some of the larger microorganisms and his statements about the doctrine of contagion are even more explicit than those of Fracastoro.

The great pioneer of modern microscopy was *Anthoni van Leeuwenhoek* (1632-1723), a Dutch linen draper, who ground his own lenses and made hundreds of microscopes. Few of his instruments provided a magnification of more than 160X but he is generally credited as making the first observations of germs, reported in his communications to the Royal Society in London. Leeuwenhoek was the first to describe spermatozoa and he gave the first complete account of the red blood corpuscles in 1674; he also found that the film from his own teeth contained "little animals, more numerous than all the people in the Netherlands."

In 1847, *Ignaz Philipp Semmelweis* (1818-1865), an assistant at the Vienna General Hospital in the maternity clinic (the world's largest at the time), was investigating the 29% postpartum mortality rate among women in Ward One, where births were handled by medical students, vs. a 3% rate in Ward Two where births were handled by midwifery pupils. Semmelweis noticed that the appearances of these deaths from puerperal fever looked the same as those observed in the body of an older colleague, forensic medicine Professor *Jakob Kolletschka* (1803-1847), who had died from adissection wound suffered in the clinic. He correctly surmised that the postpartum deaths were caused by infection from "putrid particles" carried on the hands of medical students who often shuttled back and forth between the labour wards, the obstetrical clinic, and the autopsy room where dissections were performed. Semmelweis instituted a simple routine of hand-washing with water solutions of chloride of lime, which promptly reduced mortality to 1.27%. Unfortunately, his ideas were met with fierce opposition by the conservative medical community in Vienna, who subjected him to laughter and ridicule. In disgust, Semmelweis left Austria for Budapest. There he became head of the obstetrical division of St. Rochus Hospital, where puerperal fever mortality rates were reduced to below 1% after Semmelweis introduced chlorinated water disinfection.

The man who elucidated the true nature of infection, founded the science of bacteriology, and paved the way for Lister and the antiseptic system in surgery was *Louis Pasteur* (1822-1895). Pasteur was led to his great discoveries regarding bacteria and other microorganisms by his investigations into the process of fermentation. He showed conclusively that fermentation was brought about by some external agent entering the wine. He proved by painstaking experiments under rigorously controlled conditions that meat and fluids like blood did not putrefy if they were kept in such a way that all air was excluded from them. By taking samples of air at different levels Pasteur showed that contamination became less with increasing altitude. Then he proved that the contaminating agents were living organisms (bacteria) which were everywhere—in every room, in the air, on every article of clothing, on furniture, on the ground, and on the skin. He showed that putrefaction was caused by the presence of bacteria and that this applied to putrefaction in foods (milk, wine, and meat), urine, and in wounds.

The application of Pasteur's discoveries to surgical practice was the work of *Joseph Lister* (1827-1912), a young English surgeon who had concluded that it must also be bacteria that caused the suppuration, pus and gangrene which plagued the surgical wards of those days. He determined to prevent the access of organisms by killing them in or on the surface of the wound. Pasteur had shown that heat could kill microbes, but it was impossible to apply heat to a wound without burning the patient, so some chemical substance had to be found. After trying various chemical agents he finally selected carbolic acid, and he insisted that everything which touched the wound, the dressings, the instruments and the fingers, should be treated with this *antiseptic*. He even produced an antiseptic atmosphere by means of a carbolic spray. The clinical results of Lister's first antiseptic system in 1865 included 11 compound fracture cases with only one death, a 9% mortality rate, marking a watershed between the primitive and modern eras of surgery.

Throughout the 19th century, the arguments continued as to whether microorganisms seen in a sick patient were merely coincidental with the illness, or resulted from the changes brought about by the illness itself. In 1882 the German microbiologist *Robert Koch* (1843-1910) formulated three famous postulates to guide scientists searching for disease-causing microbes. Koch argued that to prove an organism causes a disease, microbiologists must show that the organism occurs in every case of the disease; that it is never found as a harmless parasite associated with another disease; and that once the organism is isolated from the body and grown in laboratory culture, it can be introduced into a new host and produce the disease again. (An oft-stated fourth postulate, that the microbe must be isolated again from the second host, was not part of Koch's original formulation.) Koch and his pupils discovered specific bacillary causes for various diseases, including anthrax, cholera, tuberculosis, gonorrhea, diphtheria, leprosy, typhoid, trypanosomiasis, and malaria.

The theory that specific germs could cause specific disease remained contentious until the beginning of the 20th century. Many scientists of great repute rejected Koch's conclusions, with one scientist confidently asserting that "no microbe found in the living blood of any animal was pathogenic." In one celebrated case, *Max von Pettenkofer* of Bavaria (1818-1901), a distinguished 19th century experimental hygienist, induced Koch to send him a sample of his cholera vibrios culture and then wrote a letter back to Koch in 1892, as follows:

"Herr Doctor Pettenkofer presents his compliments to Herr Doctor Professor Koch and thanks him for the flask containing the so-called cholera vibrios, which he was kind enough to send. Herr Doctor Pettenkofer has now drunk the entire contents and is happy to be able to inform Herr Doctor Professor Koch that he remains in his usual good health."

Apparently Pettenkofer, aged 74 at the time he wrote the letter, survived this cholera exposure quite well, perhaps possessing the high stomach acidity which sometimes neutralizes the bacillus, though he shot himself to death in Munich 9 years later.

Skeptics notwithstanding, microbes were key. Enormous new vistas now lay open, as surgeons could confidently make an incision through intact skin without incurring an extreme risk of wound infection. The next step was to progress beyond killing wound bacteria with chemical antiseptics, to the prevention of bacterial contamination by eliminating bacteria in the operating theater—aseptic surgery. The use of steam sterilization of instruments, dressings and gowns, the wearing of masks, caps and gloves, air filtration and the other rituals of the operating theater of today were introduced over the decades following Lister's efforts.

Anesthesia and antisepsis enabled surgeons to carry out procedures that had formerly been quite beyond them, including long operations inside the head, the abdomen and the pelvis. *Theodor Billroth* of Vienna (1829-1894) resected the esophagus in 1872, parts of the intestines in 1878, and the pyloric end of the stomach in 1881. Billroth also made the first complete excision of the larynx. The first successful repair of a gunshot wound on a major artery was performed in Chicago by *John B. Murphy* (1857-1916) in 1897; *Ludwig Rehn* (1849-1930) of Frankfurt am Main performed the first successful repair of a cardiac injury in Germany in the same year.

Writing in 1874, *Sir John Eric Erichsen* (1818-1896), Professor of Surgery at University College, London, had predicted that "the abdomen, the chest, and the brain would be forever shut from the intrusions of the wise and humane surgeon." By the time of Erichsen's death 22 years later, surgeons had successfully removed from patients the stomach and large parts of the intestines, a whole lung had been excised, and a brain tumour had been extirpated. These operations were not mere feats of surgical showmanship; they saved the lives and restored the health of thousands of human beings.

Cells and tissues

The doctrine of the essential cellular nature of living things was established by 1840. Modern cell theory began in botany. The Jena botanist *Matthias Schleiden* (1804-1881) observed that plants were aggregates of cells, existing as self-reproducing living units. Exploring analogies between animals and plants in structure and growth, Theodor *Schwann* (1810-1882) took up the idea, maintaining that all these phenomena could also be demonstrated in animal structures. Thus living cells were basic to living things, and cells incorporated a nucleus and an outer membrane.

Jacob Henle (1809-85) applied cell biology to man. His three-volume *Handbuch der systematischem Anatomie des Menschen (Handbook of Systematic Human Anatomy)* (1866-1871) addressed the body from an architectural standpoint, describing its macro- and microscopic structure. Henle discovered kidney tubules and was the first to describe the muscular coat of the arteries, the minute anatomy of the eye and various skin structures, earning him a reputation as the Vesalius of histology (the study of tissues). Physiologists stressed organ and tissue function—*Claude Bernard* (1813-1878) emphasized the experimental method in establishing biological knowledge and urged that medical practice should be grounded in such knowledge.

Histology was raised to the status of an independent science by the Swiss microanatomist *Albert von Kolliker* (1817-1905), who wrote the first textbook on the subject, *Handbuch der Gewebelehre des Menschen (Handbook of the Tissues of Man)* (1852). The medical implications of cell theory were taken up by *Rudolph Virchow* (1821-1902), who dominated German biomedical research for half a century. Virchow extended the cell concept to diseased tissues; his *Die Cellularpathologie (Cellular Pathology)* (1858) analyzed such tissues from the point of view of cell formation and cell structure. Virchow initiated the idea that the body may be regarded as a "cell state in which every cell is a citizen." Disease is often but civil war, and white cells are likened to scavengers or police. Virchow maintained that cells always arose from pre-existing cells through cellular division. Since Kolliker and Virchow, the study of the intimate structure and workings of the cells themselves, as distinct from the tissues, has become a separate science, cytology, further extended to the study of cells in disease, or cytopathology.

Blood transfusions

Vague references to blood transfusion are found in medieval writings, though physicians historically were far more concerned with taking blood out of the body than with putting it back in. For example, one story tells of an attempt to prolong the life of Pope Innocent VIII (d. 1492) by means of a blood transfusion. By one account, a Jewish physician transfused the aged Pontiff with blood from three small boys, who each received one ducat as their reward; but another account says the blood was drunk, not infused. One of the earliest proposals to transfuse blood was by *Andreas Libavius* (1540-1616), a physician of Halle in Saxony, in 1615.

The first serious attempts at blood transfusion were made in England and France. In 1657 *Sir Christopher Wren* (1632-1723) carried out numerous experiments on the injection of liquids into the veins of animals. *Richard Lower* (1631-1691) of Oxford carried out the first successful transfusion from artery to vein between two dogs in 1665, a feat repeated by the French physician *Jean-Baptiste Denys* (1625-1704), who went on to attempt transfusion of blood from one kind of animal into another. Finally, before the Royal Society in 1667, Lower transfused a human youth, 15 years of age, from a sheep. The patient was greatly improved and the only ill effect was a feeling of great heat along his arm; subsequent patients were not so lucky. Blood transfusions from lambs and calves to humans continued to be tried in the mid 17th century, but were not very successful, and in 1670 were finally forbidden by law in England.

The first known transfusion of human blood into an already moribund person was attempted by *James Blundell* in 1818, and on several other occasions during the 1820s, with poor results. Transfusion was carried out on a small scale during the American Civil War, but technical difficulties connected with premature clotting and the occurrence of accidents arising from the use of incompatible blood prevented the rapid acceptance of the process. The cause of many of the untoward effects of blood transfusion was finally explained in 1901 when the presence of agglutinins and iso-agglutinins in the blood was demonstrated by *Karl Landsteiner* (1868-1943) in Vienna, winning him the Nobel Prize in 1930; in 1907, the four main blood groups were determined by *Jan Jansky* (1873-1921) of Prague. *Rhesus factor* was later identified by two American scientists, *Philip Levine* (1900-1987) and *Rufus Stetson* (1886-1967). These advances were of fundamental importance and it became possible in the 20th century to eliminate most of the fatalities due to incompatibility, allowing blood transfusions to be practiced reliably and with uniformly good results.

20th century medicine

The 20th century saw more discoveries and advances in medical science than all previous centuries combined. In this period, medicine became more powerful than ever before as scientists gained knowledge of matters and processes of illness that, at the beginning of the century, were still unknown or mysterious. Unlike the mere palliatives of earlier eras, 20th century physicians could actually cure some diseases, reverse some physical traumas, and save many lives that could not be saved before.

In the first half of the 20th century, the rational scientific paradigm that arose in the 19th century was pursued and extended. Acceptance of the germ theory of infection and the discovery of leukocytes led to the rapid emergence of immunology. This allowed medical scientists to produce protective vaccines and antisera which largely eliminated many diseases that were previously prevalent and dangerous, including whooping cough, measles, and diphtheria—the first truly effective medical treatments. Acceptance of the germ theory also led to the discovery of filterable viruses (organisms that could pass through the pores of all known filters) in the 1890s, and subsequently to the identification of viruses as specific causes of yellow fever, smallpox, typhus (a Rickettsia organism), measles, poliomyelitis, rabies, and viral meningitis. Biochemists synthesized vitamins which were recognized as essential constituents of a healthy diet, thus allowing the elimination of vitamin deficiency diseases such as scurvy, rickets, osteomalacia, beriberi, pellagra, xerophthalmia, nyctalopia, and pernicious anemia, via dietary supplements. Many metabolic diseases became treatable due to biochemical investigations; for example, the discovery of insulin in 1921 by the Canadian physiologists Sir *Frederick Banting* (1891-1941) and *Charles Best* (1899-1978) rapidly transformed diabetes from an invariably and often rapidly fatal disease into one that could be at least partially controlled, allowing sufferers many years of good life.

Blood group specification made transfusions convenient, facilitating dramatic advances in many branches of medicine,especially surgery. The first report of a successful autotransplant of a kidney into the neck of a dog was performed by *Emerich Ullmann* (1861-1937) in 1902. The first successful kidney graft between identical twins was performed in 1954 by *Joseph E. Murray* (b. 1919), who received a Nobel Prize for his work. The first heart transplantation in man was achieved by *Christiaan Barnard* (b. 1922) in 1967. Results were poor at first, but the use of new anti-rejection techniques such as the drug cyclosporin in the late 1970s, aided by advances in immunology, greatly improved the success of this and other organ transplants. By 1987, a total of ~7000 human hearts had been transplanted. At the close of the 20th century, heart, lung, heart-lung, and liver transplants were standard procedures, while grafts of small intestine and pancreas (the latter for refractory diabetes) were under active clinical investigation.

Two pivotal events transformed scientific medicine from a merely rational basis to a molecular basis, thus laying the groundwork for 21st century nanomedicine. The first pivotal event was the drug revolution, among which the most useful and spectacular were the antibiotics introduced between 1935-1945 and widely used ever since. *Antibiotics* are significant because they actively interfere with microbial metabolism and growth at the molecular level. The first antibiotic drugs were the *sulphonamides*, in 1935. Then penicillin became available in the 1940s, initially in very small quantities, then mass-produced by Pfizer during and after World War II, as a result of the research work of *Alexander Fleming* (1888-1955), *Howard Florey* (1898-1968) and *Ernst Chain* (1906-1979). In 1943, *Selman A. Waksman* (1888-1973) discovered streptomycin, the first effective anti-tuberculosis drug, for which he received the 1952 Nobel Prize. For the first time, physicians had true cures for many diseases,especially the most common bacterial diseases. Antifungal, anti-parasitic, and antiviral drugs of more limited effectiveness soon followed.

The 20th century also produced drugs that altered mood and levels of consciousness. *Barbiturates* were first introduced in 1903 (e.g. barbitone or Veronal), followed by phenobarbitone (Luminal) in 1912 and *Evipan*, the barbiturate anesthetic, in 1932. By mid-century these highly addictive drugs began to be replaced by the somewhat less-addictive *benzodiazepines*, including *Valium* and *Librium*. Tranquilizers, largely the phenothiazines such as *chlorpromazine* (*Thorazine*) and antimanics such as *lithium carbonate*, came to be widely used in psychiatry as effective medications for major mental illnesses including schizophrenia and manic depression.

The second pivotal event was the genetics revolution, starting with the discovery in 1953 of the information-carrying double-helix structure of DNA by *Francis Crick* (b. 1916) and *John B. Watson* (b. 1928), followed in the 1980s by the ability to chemically read the genetic code, isolate specific genes and clone them for further study. In the mid-1980s, the Human Genome Project was launched, with the objective of fully sequencing every gene in the human genome. The first phase of this project neared completion as the 20th century drew to a close.

Molecular biology became the premier scientific discipline of the latter 20th century. By 1998, the compositions of organs, tissues, cells, organelles, and membranes had been defined, and the biosynthesis and catabolism of hundreds of compounds had been elucidated. The regulation of body processes was described at progressively finer levels in biochemical language. Many pharmacologic agents were finally understood in terms of specific molecular loci and mechanisms of action. Advances were particularly rapid inimmunology, virology, cellular biology, peptide research, and structural biology. A beginning was made in explaining human behaviour in mechanistic terms, as more and more chemical mediators and pharmacologic modifiers were discovered. In biology, these disciplines developed porous boundaries with related disciplines such as physiology, pharmacology, neurosciences, biochemistry and biophysics. All entered a phase of confluence, employing the common language of chemistry.

Thus the late 20th century is best regarded as the molecular age of basic biological science. The molecular influence pervades all the traditional disciplines underlying clinical medicine. As of February 1999, one source listed exactly 1446 genetic disorders, most of which could be linked to a specific human chromosome; another source stated that ~4000 genetic disorders were known in 1998. There were more than 575 known abnormal human hemoglobins, and for each of these the precise structural defect in the DNA of the mutant gene could be defined. Knowledge of membrane, cytoplasmic, and nuclear receptors for hormones and drugs was exploding, with old as well as new diseases being defined in terms of receptor abnormalities—for example, type II hyper-cholesterolemia and nephrogenic diabetes insipidus. Recognition of opiate receptors led to the discovery of endogenous peptides (endorphins) with analgesic activity. Their localization promised further understanding of the limbic system, affective states, and addictions.

Defects in a subcellular organelle, the peroxisome, were known to be responsible for a growing list of important genetic afflictions such as Refsum disease, Zellweger syndrome, and X-linked adrenoleukodystrophy. The genetic defect responsible for Huntington's disease was discovered, and in the 1990s the genetic defects responsible for many other important neurological disorders were becoming known, including forms of Alzheimer's disease, Charcot-Marie-Tooth disease, and common blinding retinal degenerative disorders such as retinitis pigmentosa, Leber's hereditary optic neuropathy, Norrie's disease, and choroideremia.

DNA sequencing techniques and restriction endonucleases permitted precise identification of the exact structural alteration of the gene in an increasing number of hereditary diseases. For example, Burkitt's lymphoma is characterized by a translocation of the distal end of the long arm of chromosome 8 to loci on chromosomes 14, 22, or 2. Gene therapy—both pharmacologic modification of specific gene action and physical replacement of damaged genetic segments—became possible in experimental systems. Complete maps of the entire genomes of 18 microbial species had been compiled and published by the end of 1998, with more than 60 others in progress. In 1992, wrote the conservative *Cecil Textbook of Medicine*: "The expansion of the knowledge bank of the past quarter century justifies great optimism for the eventual control and cure of major diseases and the possible elimination of premature death from illness."

Global changes in progressive aging dysfunction were shown to be strongly related to declining secretion of growth hormone by Rudman in 1990. Shortly thereafter, anti-aging medicine was recognized as a distinct discipline and was promoted by the American Academy of Anti-Aging Medicine (A^4M), a group that claimed >6000 physician and scientist members worldwide by 1998 and had held numerous conferences. In a book on the subject, *Ronald Klatz*, A^4M's president, first comprehensively documented the implications of Rudman's work.

While biological science was vaulting forward into the molecular realm at a blistering pace, biomedical engineering lagged considerably behind, though many remarkable successes had been achieved since mid-century. Radiology expanded with sophisticated radiotherapy, ultrasound, scanning and imaging techniques (e.g., CAT, PET, NMR), with submillimeter resolution in living tissues. Surgical instruments became less damaging and less invasive; minimally invasive techniques used OK (orifice and keyhole) surgery and surgery performed under the "eye" of a scan. Fetuses could be screened for abnormality and surgery then performed upon them inside the womb, or they could be removed temporarily from the womb and then returned to continue gestating, after surgery. Prospective parents with fertility difficulties could make use of a wide range of therapies including *in vitro* fertilization. Implantable pacemakers, defibrillators, and ventricular assist devices were commonplace, and in 1998 artificial wearable/implantable and full/partial replacements for lungs, heart, kidney, liver, and pancreas were either available, in clinical trials, or under development. Electron microscopes with 200,000X magnifications allowed many details of internal cellular and viral structures to be resolved as early as 1946; by 1998, atomic-force (AFM) and scanning-tunneling microscopy (STM) permitted direct tactile examination of individual biomolecules in fixed cells.

As the 20th century drew to a close, a few preliminary efforts had been made to apply the molecular approach to medical diagnostic and clinical tools. Biotechnology was one avenue being pursued, with the rational design of artificial enzymes and specified-ligand binding sites having already been achieved in certain limited cases, and the beginnings of gene therapy as noted earlier. Carbon fullerenes had been used to create a water-soluble inhibitor of HIV protease, and had other biological applications. AFM-based force-amplified biological sensors could detect defined biological species such as cells, proteins, toxins, and DNA at concentrations as low as 10^{-18} M ($\sim 1/mm^3$), and automated laboratory systems for sorting and handling individual cells and viruses were commonplace. Biotech companies such as Physiome Sciences of Princeton NJ had developed three-dimensional computer models of the heart and other organs. Physiome's heart model was based on detailed molecular, biochemical, cellular and anatomical information, including submodels of all the different cell types found in the heart embodying knowledge of the function of each cell type in healthy and diseased hearts, and information on gene function and the causes and effects of congestive heart failure, arrhythmias and heart attacks. Other cell biochemistry simulators included E-CELL and the Virtual Cell.

There was also much progress and interest in nanostructure analysis and nanomaterials fabrication for medical and biological purposes, with research groups emerging at major universities such as the Cornell Nanofabrication Facility, theUniversity of Michigan Center for Biologic Nanotechnology, the Rice University Center for Nanoscale Science and Technology, the CalTech Materials and Process Simulation Center, the Washington University Nanotechnology Center (St. Louis, MO), the USC Laboratory for Molecular Robotics, the UCLA Exotic Materials Center, the Institute for Molecular Medicine of the University of Oxford, and at many biotechnology-oriented corporations such as Nanogen and Affymetrix. However, the greatest medical revolution of all awaits the ability to engineer and fabricate whole devices and systems at the molecular scale. Along this course lies nanotechnology and molecular manufacturing, a deep well from which nanomedicine will inevitably spring.

21st century medicine

It is always somewhat presumptuous to attempt to predict the future, but in this case we are on solid ground because most of the prerequisite historical processes are already in motion and all of them appear to be clearly pointing in the same direction.

Medical historian *Roy Porter* notes that the 19th century saw the establishment of what we think of as scientific medicine. From about the middle of that century the textbooks and the attitudes they reveal are recognizable as not being very different from modern ones. Before that, medical books were clearly written to address a different mind-set.

But human health is fundamentally biological, and biology is fundamentally molecular. As a result, throughout the 20th century scientific medicine began its transformation from a merely rational basis to a fully molecular basis. First, antibiotics that interfered with pathogens at the molecular level were introduced. Next, the ongoing revolutions in genomics, proteomics and bioinformatics provided detailed and precise knowledge of the workings of the human body at the molecular level. Our understanding of life advanced from organs, to tissues, to cells, and finally to molecules, in the 20th century. By the early 21st century, the entire human genome will be mapped. This map will inferentially incorporate a complete catalog of all human proteins, lipids, carbohydrates, nucleoproteins and other molecules, including full sequence, structure, and much functional information. Only some systemic functional knowledge, particularly neurological, may still be lacking by that time.

This deep molecular familiarity with the human body, along with simultaneous nanotechnological engineering advances, will set the stage for a shift from today's molecular scientific medicine in which fundamental new discoveries are constantly being made, to a molecular technologic medicine in which the molecular basis of life, by then well-known, is manipulated to produce specific desired results. The comprehensive knowledge of human molecular structure so painstakingly acquired during the 20th and early 21st centuries will be used in the 21st century to design medically-active microscopic machines. These machines, rather than being tasked primarily with voyages of pure discovery, will instead most often be sent on missions of cellular inspection, repair, and reconstruction. In the coming century, the principal focus will shift from medical science to medical engineering. Nanomedicine will involve designing and building a vast proliferation of incredibly efficacious molecular devices, and then deploying these devices in patients to establish and maintain a continuous state of human healthiness.

The very earliest nanotechnology-based biomedical systems may be used to help resolve many difficult scientific questions that remain. They may also be employed to assist in the brute-force analysis of the most difficult three-dimensional structures among the 100,000-odd proteins of which the human body is comprised, or to help ascertain the precise function of each such protein. But much of this effort should be complete within the next 20-30 years because the reference human body has a finite parts list, and these parts are already being sequenced, geometered and archived at an ever-increasing pace. Once these parts are known, then the reference human being as a biological system is at least physically specified to completeness at the molecular level. Thereafter, nanotechnology-based discovery will consist principally of examining a particular sick or injured patient to determine how he or she deviates from molecular reference structures, with the physician then interpreting these deviations in light of their possible contribution to, or detraction from, the general health and the explicit preferences of the patient.

In brief, nanomedicine will employ molecular machine systems to address medical problems, and will use molecular knowledge to maintain human health at the molecular scale.

Volitional Normative Model of Disease

What, exactly, is "medicine"? Dictionaries give several definitions, ranging from the very restrictive to the most general, as follows: "a drug or remedy"; "any substance used for treating disease"; "any drug or other substance used in treating disease, healing, or relieving pain"; "in a restricted sense, that branch of the healing art dealing with internal diseases"; "treatment of disease by medical, as distinguished from surgical, treatment"; "the branch of this science and art that makes use of drugs, diet, etc., as distinguished especially from surgery and obstetrics"; "the study and treatment of general diseases or those affecting the internal parts of the body"; "the science of treating disease, the healing art"; "the art and science of preventing or curing disease"; "the act of maintenance of health, and prevention and treatment of disease and illness"; "the department of knowledge and practice dealing with disease and its treatment"; or, most generally, "the science and art of diagnosing, treating, curing, and preventing disease, relieving pain, and improving and preserving health". In this book, we shall adopt the latter, maximally-inclusive, definition of "medicine".

The contemporary physician might at first be inclined to relegate molecular approaches to some minor subfield, perhaps "nanoanalytics," "nanogenomics," or "nanotherapeutics." This would be a serious mistake, because the application of molecular approaches to health care will significantly impact virtually every category of laboratory and clinical practice across the board. Thus we are led to the broadest possible conception of nanomedicine as "the science and technology of diagnosing, treating, and preventing disease and traumatic injury, of relieving pain, and of preserving and improving human health, using molecular tools and molecular knowledge of the human body."

This brings us to the question of "disease," a complex term whose meaning is still hotly debated among medical academics. Illnesses due to microorganisms, or conditions in which the doctor's contribution to the diagnosis was important, were most likely to be called a disease, but if the cause was a known physical or chemical agent the condition was less likely to be regarded as disease; general practitioners also had the broadest definition of disease. No less than eight different types of disease concepts are held by at least some people currently engaging in clinical reasoning and practice, including:

Disease nominalism

A disease is whatever physicians say is a disease. This approach avoids understanding and forestalls inquiry, rather than furthering it.

Disease relativism

A disease is identified or labeled in accordance with explicit or implicit social norms and values at a particular time. In 19th century Japan, for example, armpit odor was considered a disease and its treatment constituted a medical specialty. Similarly, 19th-century Western culture regarded masturbation as a disease, and in the 18th century, some conveniently identified a disease called drapetomania, the"abnormally strong and irrational desire of a slave to be free."

Various non-Western cultures having widespread parasitic infection may consider the lack of infection to be abnormal, thus not regarding those who are infected as suffering from disease.

Sociocultural disease

Societies may possess a concept of disease that differs from the concepts of other societies, but the concept may also differ from that held by medical practitioners within the society itself. For instance, hypercholesterolemia is regarded as a disease condition by doctors but not by the lay public; medical treatment may be justified, but persons with hypercholesterolemia may not seek treatment, even when told of the condition. Conversely, there may be sociocultural pressure to recognize a particular condition as a disease requiring treatment, such as alcoholism and gambling.

Statistical disease

A condition is a disease when it is abnormal, where abnormal is defined as a specific deviation from a statistically-defined norm. This approach has many flaws. For example, a statistical concept makes it impossible to regard an entire population as having a disease. Thus tooth decay, which is virtually universal in humans, is not abnormal; those lacking it are abnormal, thus are "diseased" by this definition. More reasonably, a future highly-aseptic society might regard bacterium-infested 20th century humans (who contain in their bodies more foreign microbes than native cells) as massively infected. Another flaw is that many statistical measurables such as body temperature and blood pressure are continuous variables with bell-shaped distributions, so cutoff thresholds between "normal" and "abnormal" seem highly arbitrary.

Infectious agency

Disease is caused by a microbial infectious agent. Besides excluding systemic failures of bodily systems, this view is unsatisfactory because the same agent can produce very different illnesses. For instance, infection with hemolytic *Streptococcus* can produce diseases as different as erysipelas and puerperal fever, and Epstein-Barr virus is implicated in diseases as varied as Burkitt's lymphoma, glandular fever, and naso-pharyngeal carcinoma.

Disease realism

Diseases have a real, substantial existence regardless of social norms and values, and exist independent of whether they are discovered, named, recognized, classified, or diagnosed. Diseases are not inventions and may be identified with the operations of biological systems, providing a reductionistic account of diseases in terms of system components and subprocesses, even down to the molecular level. One major problem with this view is that theories may change over time —almost every 19th century scientific theory was either rejected or highly modified in the 20th century. If the identification of disease is connected with theories, then a change in theories may alter what is viewed as a disease. For example, the 19th century obsession with constipation was reflected in the disease labelled "autointoxication," in which the contents of the large bowel were believed to poison the body. Consequently much unnecessary attention was paid to laxatives and purgatives and, when surgery of the abdomen became possible toward the end of the century, operations to remove the colon became fashionable in both England and America.

Disease idealism

Disease is the lack of health, where health is characterized as the optimum functioning of biological systems. Every real system inevitably falls short of the optimum in its actual functioning. But by comparing large numbers of systems, we can formulate standards that a particular system ought to satisfy, in order to be the best of its kind. Thus "health" becomes a kind of Platonic ideal that real organisms approximate, and everyone is a less than perfect physical specimen. Since we are all flawed to some extent, disease is a matter of degree, a more or less extreme variation from the normative ideal of perfect functioning. This could be combined with the statistical approach, thus characterizing disease as a statistical variation from the ideal. But this view, like the statistical, suffers from arbitrary thresholds that must be drawn to qualify a measurable function as representing a diseased condition.

Functional failure

Organisms and the cells that constitute them are complex organized systems that display phenomena (e.g. homeostasis) resulting from acting upon a program of information. Programs acquired and developed during evolution, encoded in DNA, control the processes of the system. Through biomedical research, we write out the program of a process as an explicit set (or network) of instructions. There are completely self-contained "closed" genetic programs, and there are "open" genetic programs that require an interaction between the programmed system and the environment, e.g. learning or conditioning. Normal functioning is thus the operation of biologically programmed processes, e.g. natural functioning, and disease may be characterized as the failure of normal functioning. One difficulty with this view is that it enshrines the natural as the benchmark of health, but it is difficult to regard as diseased a natural brunette who has dyed her hair blonde in contravention of the natural program, and it is quite reasonable to regard the mere possession of an appendix as a disease condition, even though the natural program operates so as to perpetuate this troublesome organ. A second weakness of this view is that disease is still defined against population norms of functionality, ignoring individual differences. As a perhaps overly simplistic example, 65% of all patients employ a cisterna chyli in their lower thoracic lymph duct, while 35% have no cisterna chyli—which group has a healthy natural program, and which group is "diseased"?

A ninth view of disease is also proposed, a new alternative which seems most suitable for the nanomedical paradigm, called the "volitional normative" model of disease. As in the "*disease idealism*" view, the volitional normative model accepts the premise that health is the optimal functioning of biological systems. Like the "functional failure" view, the volitional normative model assumes that optimal functioning involves the operation of biologically programmed processes.

However, two important distinctions from these previous views must be made. First, in the volitional normative model, normal functioning is defined as the optimal operation of biologically programmed processes as reflected in the patient's own individual genetic instructions, rather than of those processes which might be reflected in a generalized population average or "Platonic ideal" of such instructions; the relative function of other members of the human population is no longer determinative. Second, physical condition is regarded as a volitional state, in which the patient's desires are a crucial element in the definition of health. This is a continuation of the current trend in which patients frequently see themselves as active partners in their own care.

In the volitional normative model, disease is characterized not just as the failure of "optimal" functioning, but rather as the failure of either (a) "optimal" functioning or (b) "desired" functioning. Thus disease may result from:

1. A failure to correctly specify desired bodily function (specification error by the patient);
2. A flawed biological program design that doesn't meet the specifications (programming design error);
3. Flawed execution of the biological program (execution error);

4. External interference by disease agents with the design or execution of the biological program (exogenous error); or
5. Traumatic injury or accident (structural failure).

In the early years of nanomedicine, volitional physical states will customarily reflect "default" values which may differ only insignificantly from the patient's original or natural biological programming. With a more mature nanomedicine, the patient may gain the ability to substitute alternative natural programs for many of his original natural programs. For example, the genes responsible for appendix morphology or for sickle cell expression might be replaced with genes that encode other phenotypes, such as the phenotype of an appendix-free cecum or a phenotype for statistically typical human erythrocytes. Many persons will go further, electing an artificial genetic structure which, say, eliminates age-related diminution of the secretion of human growth hormone and other essential endocrines. (The graduated secretion of powerful proteolytic enzymes, perhaps targeted for gene-expression in appropriate organs, may reverse and control the accumulation of highly crosslinked collagenaceous debris; by 1998, many members of the mainstream medical community were already starting to regard aging as a treatable condition.) On the other hand, a congenitally blind patient might desire, for whatever personal reasons, to retain his blindness. Hence his genetic programs that result in the blindness phenotype would not, for him, constitute "disease" as long as he fully understands the options and outcomes that are available to him. (Retaining his blindness while lacking such understanding might constitute a specification error, and such a patient might then be considered "diseased.") Whether the broad pool of volitional human phenotypes will tend to converge or diverge is unknown, although the most likely outcome is probably a population distribution (of human biological programs) with a tall, narrow central peak (e.g., a smaller standard deviation) but with longer tails (e.g., exhibiting a small number of more extreme outliers).

One minor flaw in the volitional normative model of disease is that it relies upon the ability of patients to make fully informed decisions concerning their own physical state. The model crucially involves desires and beliefs, which can be irrational, especially during mental illness, and people normally vary in their ability to acquire and digest information. Patients also may be unconscious or too young, whereupon default standards might be substituted in some cases. Nevertheless, the volitional normative view of disease appears most appropriate for nanomedicine because it recognizes that the era of molecular control of biology could bring considerable molecular diversity among the human population. Conditions representing a diseased state must of necessity become more idiosyncratic, and may progressively vary as personal preferences evolve over time. Some patients will be more venturesome than others— "to each his own." As an imperfect analogy, consider a group of individuals who each take their automobile to a mechanic. One driver insists on having the carburetion and timing adjusted for maximum performance (the "racer"); another driver prefers optimum gas mileage (the "cheapskate"); still another prefers minimizing tailpipe emissions (the "environmentalist"); and yet another requires only that the engine be painted blue (the "aesthete"). In like manner, different people will choose different personal specifications. One can only hope that the physician will never become a mere mechanic even in an era of near-perfect human structural and functional information; an automobile conveys a body, but the human body conveys the soul. Agrees theorist Guttentag: "The physician-patient relationship is ontologically different from that of a maintenance engineer to a machine or a veterinarian to an animal."

Treatment Methodology

The availability of advanced nanomedical instrumentalities should not significantly alter the classical medical treatment methodology, although the patient experiences and outcomes will be greatly improved. Treatment in the nanomedical era will become faster and more accurate, efficient, and effective. In clinical practice, patient treatment customarily includes up to six distinguishable phases: examination, diagnosis, prognosis, treatment, validation, and prophylaxis. Let us consider each of these, in turn.

Examination

The first step in any treatment process is the examination of the patient, including the individual's medical history, personal functional and structural baseline, and current complaints. In classical medicine, interview and observation have long been the cornerstone of examination. In ancient times this was limited to obvious manifestations and simple constellations of observables, such as the Hippocratic facies, the four signs of inflammation noted by Celsus, or pulse rate and fever. Clinicians recognize that the traditional taking and interpreting of oral medical histories from new patients is a subtle and complex art, although some aspects of this process might be automated using voice recognition and text preinterpretation software, somewhat easing the physician's burden. Advancing technology has also brought a plethora of tests that contribute to accurate diagnosis, including auscultation, microscopy and clinical bacteriology in the 19th century, and radiological scanning, clinical biochemistry, genetic testing, and minimally invasive exploratory surgery in the 20th century.

In the 21st century, new tools for nanomedical testing and observation will include clinical *in vivo* cytography; real-time whole-body microbiotic surveys; immediate access to laboratory-quality data on the patient (e.g. blood tests such as blood counts, dissolved gases and solutes, vitamin and ion assays); physiological function and challenge tests; tissue composition including direct organelle counts in specified tissue populations; quantitative flowcharts of *in cyto* secondary messenger molecules, extracellular hormones and neuropeptides; per-compartment cytoglucose inventories; and so forth. Before a proper diagnosis can be made, the physician must also establish the patient's personal functional and structural baseline against which any deviations can be noted and corrected, in keeping with the volitional normative model of disease.

The capabilities of nanomedial testing are explored at length. By way of introduction, it is instructive to think about a trivial class of test procedures that might be used to diagnose a simple infectious disease at several different levels of technological competence. Let us consider a patient who presents with signs and symptoms that are nonspecific in nature but which suggest an infectious process— e.g. nasal congestion, mild fever, discomfort and cough. The initial signs are due in part to the body's inflammatory response and in part to the infectious agent itself. The diagnostic goal is to identify the infectious agent.

In the late 20th century, the usual procedure would be to culture a sample taken from the patient, in the microbiology laboratory, using various broths, petri plates, and biochemical tests. Some infectious agents are easy to demonstrate. Beta hemolytic streptococci from a throat swab will grow overnight on a blood agar plate, and colony counts for *E. coli* in a urine sample are available in 24 hours. A throat culture that the lab reports as a mixed culture causes no excitement, and a single isolate of *Staphylococcus epidermidis* in a blood culture is usually regarded as a skin contaminant.

Moving up to a higher level of technological competence, biotechnologists describe an ideal diagnostic scenario which takes a molecular approach to the diagnosis of infectious disease using recombinant DNA technology. This approach was not yet possible in

1996 when first suggested, but was regarded as a reasonable and likely future application of biotechnology in the early 21st century given rapid progress in single-molecule DNA assay techniques:

"A patient presents in the clinic with mild fever, nasal congestion, discomfort, and cough. A swab of his throat is taken. Instead of culture to identify abnormal microorganisms by their pattern of growth, the sample is analyzed by recombinant DNA techniques. The cotton throat swab is mixed with a cocktail of DNA probes. Enzymes that digest and release the DNA from both host cells and invading bacteria make the DNA in the sample immediately available for hybridization to the probes. The swab is swirled in the liquid mix of the prepackaged test kit for 1 minute. The liquid is then poured through a column that separates hybridized DNA molecules (bacterial target DNA sequences bound to probe DNA) from all other debris [taking several minutes]. A chemi-luminescence detection system for the probes shows two of several possible colours indicating mixed infection. The diagnostic result, available in 10 minutes, indicates a Rhinovirus of a strain known to be epidemic in the geographic area. A significant superinfection with a penicillin-resistant streptococcus is also identified. With a definitive diagnosis, the patient is started on the appropriate antibiotic."

How might nanomedicine handle this test? In the nanomedical era, taking and analyzing microbial samples will be much simpler for the practitioner. Such analysis will be as quick and convenient as the electronic measurement of body temperature using a tympanic thermometer in a late 20th-century clinical office or hospital. The physician faces the patient and pulls from his pocket a lightweight handheld device resembling a pocket calculator. He unsnaps a self-sterilizing cordless pencil-sized probe from the side of the device and inserts the business end of the probe into the patient's opened mouth in the manner of a tongue depressor. The ramifying probe tip contains billions of nanoscale molecular assay receptors mounted on hundreds of self-guiding retractile stalks. Each receptor is sensitive to one of thousands of specific bacterial membrane or viral capsid ligands. An acoustic echolocation transceiver provides gross spatial mapping. The patient says "Ahh," and a few seconds later a three-dimensional colour-coded map of the throat area appears on the display panel that is held in the doctor's hand. A bright spot marks the exact location where the first samples are being taken. Underneath the colour map scrolls a continuously updated microflora count, listing in the leftmost column the names of the ten most numerous microbial and viral species that have been detected, key biochemical marker codes in the middle column, and measured population counts in the right column. The number counts flip up and down a bit as the physician directs probe stalks to various locations in the pharynx to obtain a representative sampling, with special attention to sores or any signs of exudate. After a few more seconds, the data for two of the bacterial species suddenly highlight in red, indicating the distinctive molecular signatures of specific toxins or pathological variants. One of these two species is a known, and unwelcome, hostile pathogen. The diagnosis is completed, the infectious agent is promptly exterminated, and a resurvey with the probe several minutes afterwards reveals no evidence of the pathogen.

Diagnosis

Diagnosis is the determination of the cause and nature of a disease in order to provide a logical basis for treatment and prognosis. Traditionally the diagnostic process begins with a thorough history taken from the patient and a relevant physical examination. Often this sufficed to make a confident diagnosis, but the cause of some illnesses remained uncertain without recourse to additional information such as blood tests or radiological examinations. Nanotechnology-based diagnosis will consist principally of examining the patient to determine how he or she deviates from autogenous reference structures and functions, and then interpreting those deviations as healthy or unhealthy for that patient.

In the 20th century, diagnoses frequently involved a high degree of uncertainty, largely due to the general lack of comprehensive molecular diagnostic tools. Thus diagnosis would be guided by statistical analyses; one branch of decision analysis, called utility analysis, even allowed the patient to participate in the decisionmaking process. When the correct decision is unclear, urges one textbook, it is well to remember time-honored Hippocratic aphorisms such as "first, do no harm" and "common things occur commonly." The eminent Canadian physician Sir William Osler (1849-1919) lamented that "errors of judgement must occur in the practice of an art which consists largely in balancing probabilities." Most doctors would prefer to understand the root cause of medical problems rather than adopt mere statistical approaches.

Nanomedical tools will vastly reduce diagnostic uncertainty. Using nanomedical instrumentalities, doctors will gain access to unprecedented amounts of information about their patients including in-office comprehensive genotyping and real-time whole-body scans for particular bacterial coat markers, tumour cell antigens, mineral deposits, suspected toxins, hormone imbalances of genetic or lifestyle origin, and other specified molecules, producing three-dimensional maps of desired targets with submillimeter spatial resolution. Embedded *in vivo* nanomedical data archives can provide onboard storage of regularly updated self-diagnostic scans, reducing to a minimum the need for symptomatic interview data from patients who may be unconscious, inarticulate, or verbose, who may have limited powers of self-analysis or self-observation and who may have forgotten, suppressed, or amplified descriptions of symptoms. Physicians do not require an exhaustive survey of the entire body of each patient to molecular detail to make a valid diagnosis. In any particular case, it is the function of the trained medical mind to quickly ascertain where and where not to look in molecular detail. But in the nanomedical era, powerful tools will be available to allow the practitioner to examine almost any portion of a patient in as much detail as desired, right down to the molecular level, with results available in seconds or minutes, and at reasonable cost.

Prognosis and treatment

Prognosis is a judgement or forecast, based upon a correct diagnosis, of the future course of a disease or injury, and of the patient's prospects for partial or full recovery. Guttentag identifies prognosis as "the predicted course of the reduced state of the patient's psychosomatic freedom of action," and treatment as "the physician's ability to intervene."

But prognosis is a function of treatment as well as disease. From the post-Hippocratic era through the 18th century, treatments were almost purely empirical and often did more harm than good. During the 19th and early 20th centuries, treatments were scientific but largely homeostatic—the medical intervention was rational but served mainly to assist the body in healing itself. Throughout the remainder of the 20th century, truly curative treatments began to rescue some patients from conditions from which their unaided bodies would not have been able to recover. Although conventional biotechnology will enable some important tissue and cellular replacement treatments by the early 21st century, nanomedicine will enable major reconstructive and restorative procedures at the tissue, cellular and molecular levels and will employ active antibiotic devices. The prognosis will almost always be good, except in cases of severe neural damage and a few other specialized circumstances. Therapeutic treatments will be selected to reverse all pathological effects of disease or injury, with a minimum of pain.

discomfort, side-effects, intrusiveness and time, and with a maximum of effectiveness, efficiency, and likelihood of success, though of course some tradeoffs will always exist. Nanomedicine also will excel in the correction of molecular defects of a kind which Nature has no predesigned tools—such as the breakdown and removal of intracellular lipofuscin (for which there appear to be no natural enzymes) and the removal of indigestible waste products which interfere with neuronal axon transport.

We may compare the therapeutic response to a simple infection at several different levels of technological competence. Consider a patient who has been diagnosed with eastern equine encephalitis, a mosquito- or tick-borne arbovirus. In the 20th century, there was no specific treatment for this disease. Care was generally supportive, with the doctor attempting to maintain the patient's heart and lung function while the infection ran its course. The prognosis was poor. There was a 50%-75% mortality rate with frequent sequelae including seizures and paralysis, especially in children.

Biotechnologists proposed a molecular approach to therapeutics using recombinant DNA technology that was not yet possible in 1996 but was regarded as likely by the early 21st century:

"A patient enters the hospital with high fever and intense headaches. A spinal fluid tap is submitted to the molecular microbiology lab. After screening for several viruses, a species of equine encephalitis virus is identified that is endemic to a location recently visited by the patient. A call to the Centers for Disease Control results in the emergency delivery of a new antiviral agent. Antisense oligonucleotides are injected into the cerebral spinal fluid. These small DNA pieces bind directly to the virus and block its further proliferation. A temporary reservoir giving access to the cerebrospinal fluid is placed and infusion of this therapeutic molecular inhibitor of the virus continues for 5 days until signs of encephalitis have passed."

The nanomedical therapy? As before, a nanomedical cure for eastern equine encephalitis may be far simpler, less painful and a great deal quicker. A single therapeutic dose consisting of ~0.1 cm^3 of isotonic saline fluid containing ~10 billion active micron-size virucidal nanodevices, a 10% volumetric nanodevice suspension, is injected into the cerebrospinal fluid. Each therapeutic nanorobot has chemical sensors that can unambiguously recognize fluidborne or *in cyto* arbovirus particles and, once recognition has occurred, destroy them and also reverse the cellular damage. A nanorobot population of this size should be able to destroy all viral particles and effect needed repairs in at most an hour, after which the devices are programmed and equipped either to eliminate themselves from the body or to be manually exfused (e.g., nanapheresis).

Validation and prophylaxis

A proper therapeutic protocol will include a procedure for follow-up to ensure that the prescribed treatment was correctly executed with good results. This step is often neglected in order to save costs and may be considered unimportant by some practitioners because approximately 80%-90% of all illnesses which take patients to the doctor are self-curing or self-limiting. For example, the common cold, most infectious diseases and many minor injuries are problems that usually will resolve on their own even with no treatment. In these cases the purpose of treatment is not to provide a cure, but rather to speed the healing process, improve comfort, and avoid complications. Many nanomedical treatments will require supervision and will run quickly to completion, thus follow-up may come back into vogue. Validation may also be viewed as a post-treatment re-diagnosis to ensure that no disease remains present in the patient.

Prophylaxis is the prevention of disease, typically including patient education, immunization programs, amelioration of occupational hazards, and other preventive and public health measures. In a treatment environment that is rich in effective antibacterial instrumentalities, those microbes which survive will evolve to produce only modest or negligible symptoms that are insufficiently annoying to motivate a patient to seek professional therapeutic relief. It is well-known that bacteria can modify their behaviour over time. For example, syphilis had a much more fulminating course in the Middle Ages than it has in the 20th century. Some future strain of the syphilitic microbe might produce negligible symptoms, but we should still insist on its eradication because of its potential to revert to its earlier virulence if allowed to spread unchecked in a more benign form. Preventative procedures may also be needed to discover, diagnose, and treat apparently symptomless diseases, and a variety of molecular-based physiological malfunctions and structural micropathologies may require nanoscale tools in order to detect them.

With medical conditions that require ongoing supervision and adjustment, such as maintaining optimum hormone balance and minimal accumulation of molecular debris (e.g. anti-aging medicine), nanoscale monitoring stations may act as onboard cellular guidance systems, stimulating or suppressing endocrine secretion as necessary to preserve an ideal state of equilibrium. In some cases, direct manufacture of compounds not easily produced by ribosomes or other biological organelles may be required.

Evolution of Bedside Practice

The relationship between physician and patient has been evolving in response to the rapidly changing medical environment. Two of the most important trends are the decline of traditional holism (with a concomitant increase in specialization) and the rise of therapeutic customization in medical practice.

Specialization and holistic medicine

Holism is a philosophy which holds that individuals function as complete units that cannot be reduced merely to the sum of their parts. It is unarguable that a simplistic reductionist view of the patient which ignores the complex interactions among the many cells, tissues, organs, and systems constituting the human body is deeply flawed. For example, an understanding of the molecular basis of the contractile proteins of heart muscle will not alone tell us how heart muscle cells will look or act; knowing only about the parts of something is not sufficient to predict the behaviour of the whole. However, traditional concepts of holistic medicine go well beyond such basic systems philosophy—incorporating requirements for a consideration of all physical, emotional, social, spiritual, environmental and economic needs of the patient.

N. Jewson and others have decried the modern shift away from holistic medicine, which is asserted to have taken place in three historical phases in the West. The initial phase is identified as the practice of "bedside medicine," where wealthy fee-paying clients in the 17th and 18th centuries helped shape their own diagnosis and treatment by medical practitioners in an holistic manner. Aspects of the patient's emotional and spiritual life were seen as central by the practitioner in making a diagnosis, since most physical treatments were only palliative and so the doctor had to focus on nonphysical supportive measures. This frame of reference was progressively replaced during the 19th century with the trend toward "hospital medicine," wherein physicians concentrated on generic classifications of diseases that were manifested in the patient, moving doctors away from the earlier focus on the individual as a whole person. The 20th century saw the development of "laboratory medicine," which moved diagnosis and therapy even further away from the whole patient, "who came to be medically conceived as little more than a depersonalized object, comprised of a complex of cells."

A more charitable view is that physicians have increasingly specialized in treating those physical diseases for which effective treatments may be readily specified, leaving nonphysical and nontreatable issues for psychiatrists, social workers, fitness coaches, priests, lawyers, or other professionals to deal with. In the 21st century this operational specialization may become complete, since nanomedicine phenomenologically regards the human body as an intricately structured machine with trillions of complex interacting parts, with each part (and each subsystem of parts) subject to individual scrutiny, repair, and possibly replacement by artificial technological means. In this new medical cosmology, the concept of the whole patient almost completely dissolves into a data-intensive whirlwind of molecular detail at the cellular, tissue, organ, and systemic levels.

And yet, as in a bygone era, patients once again will help to shape their own diagnosis and treatment at the hands of medical practitioners who begin to apply the volitional normative model of disease in their practices. This may breathe new life into the age-old medical school dictum to "treat the patient as a person" and to "focus on the sick person" rather than exclusively on the body.

The availability of extremely powerful and transforming molecular technologies also argues for a return to the romantic and perhaps quaint concept of a single doctor taking care of a single patient. The potential for interactions among highly potent nanorobotic instrumentalities argues for diagnostic and therapeutic "gatekeeping" by a single trusted practitioner in whom strategic treatment responsibility is vested—in partnership, of course, with the patient.

Customized diagnosis and therapeutics

In science the objective is to understand the individual occurrence by means of a general law; in medical practice, knowledge of what is generally the case does not tell the physician how to treat a particular patient. Thus the problem of how to proceed from the prototypic case to the individual instance remains to be solved in a systematic manner by the practitioner.

The practitioner of "bedside medicine" in the 17th and 18th centuries had few curative and almost no customized tools at his disposal—perhaps a few dozen basic surgical techniques, performed septically, hazardously, and without anesthesia, and a drug/herbal formulary consisting of a few hundred substances. For instance, the

Edinburgh Pharmacopoeia of 1803 listed only 222 simples while the *London Pharmacopoeia* of 1809 listed fewer than 200 items, most of which had variable, uncertain, minimal, or nonspecific potency. A few more options became available to the 19th century medical practitioner, including complex surgical techniques for specific conditions that could be performed aseptically with anesthesia and a good chance for success, a somewhat broader and more efficacious pharmacopeia, vaccines targeted to several diseases, and an improving diagnostic ability. By the 20th century, the physician could prescribe from among tens of thousands of specific drugs to target specific bacterial, viral, fungal, or parasitic infections; select from among a very precise array of anesthetic agents, chosen to avoid allergic responses in particular patients; perform a wide variety of noninvasive tests and scans for diagnostic purposes to identify very specific conditions; and perform minimally invasive surgeries directed at many arbitrary tissue masses as small as 1 mm^3 in volume. The first glimmerings of personalized genetic therapies also began to appear.

With the arrival of nanomedicine in the 21st century, the treatment paradigm will complete its transition from coarse-grained, one-size-fits-all, slow-acting methods to molecularly-precise, completely customized, speedy and highly-efficacious procedures and instrumentalities. The irregular shotgun pattern of 18th century palliatives will evolve during the early 21st century into a penetrating and perfectly tailored hail of "magic bullets" each targeting an individual cell or group of cells unique to the individual patient. The 19th century herbalist John Ayrton Paris could have been describing the nanomedical future in his popular textbook *Pharmacologia*, 1840 edition, when he wrote:

"If [a physician] prescribes upon truly scientific principles, he will rarely in the course of his practice compose two formulae that shall, in every respect, be perfectly similar, for the plain reason that he will never meet with two cases exactly alike. Now let me ask what constitutes the essential difference between the true physician and his counterfeit between the philosopher and the empiric? Simply this that the latter exhibits the same medicine in every disease, however widely each may differ from the other in its symptoms and character; while the former examines, in the spirit of philosophic analysis, all the existing peculiarities of his patient, and of his discord...and then adapts with a sound discretion and with a correct judgement of his medicinal agents, such means as may best be calculated to control and correct the patient's morbid condition."

Physician-patient relationship

Many other aspects of the physician-patient relationship, especially as this relationship may evolve in the coming era of nanomedicine, are important and worthy of extensive discussion. One such issue is the obligation of both parties in the partnership to tell the truth. The patient as a fellow human being has every right to know the truth about his or her biological condition, but other considerations may enter into the fulfillment of this obligation. Patients have a quite natural anxiety about their own possible death, and it has been claimed that this anxiety implies that no one can be truly objective toward his or her own body. Guttentag observes that "telling an unwelcome truth to the unprepared is as ill-conceived as trying to hide the truth from the prepared."

In the nanomedical era, the sheer number of "truths" that may become available for disclosure will increase enormously even as the terminal prognosis becomes rare. For example, each human being is believed to possess at least 4-10 potentially serious genetic defects; up to 1% of human DNA is of exogenous viral origin, and as much as 10% of the genome consists of transposons, discrete sequences that are positionally mobile among the chromosomes. Should something be done about this, or not? What should the average patient make of the news that his physician has discovered exactly 57 submicron-scale lamellar defects scattered throughout the compact bone of the caudal epiphysis of the patient's right humerus?

In the nanomedical era, people will gain the ability to specify their own physical structure to minute detail, but many patients will not be ready, willing, or able to assume responsibility for this knowledge. Thus there is no ideal substitute for the doctor's interpretative abilities and judgements on the patient's behalf as to the personal significance of specific diagnostic information. As the great clinician Thomas Addis observed in another context: "Honesty with patients requires thought and discipline and effort."

Perhaps the single most important aspect of the physician-patient relationship, in any century, is the humanistic quality of the good doctor. The patient seeks a physician who cares about him as a person and will diagnose and prescribe in a sensitive and compassionate manner, accepting some degree of obligation to the patient. Speaking to medical students, J.C. Bennett describes the implicit social contract between doctor and patient that will still apply in the nanomedical era, as it does today:

"To receive medical care, patients must trust their bodies and their very lives to physicians, and so to be in an honest position to give medical care, physicians must earn such radical trust. Mere technical treatment of disease does not suffice. Patients must be able reasonably to believe that their physicians care about them in an extraordinarily personal way. This exchange of care for trust, while not identical to friendship or love, is equally binding. From it develops an interdependence that is far from unwholesome; rather, it potentiates care and promotes healing. Our late twentieth century sophistication and technologic orientation have too often cost us warmth, humor, and humanity, leaving us in social isolation. We do far better as professionals to err on the side of being human with our patients, than to try to play deus ex machina, the god from the machine."

Changing View of the Human Body

How does a patient regard his or her own body, and how might this most intimate of all relationships change in the nanomedical era? The so-called dualist theory of the human compound, as originally developed by Descartes and widely accepted today by the ordinary person, holds that the human being consists of two separate kinds of thing: the body and the mind or soul. The body acts as a host or receptacle for the mind. The mind, often called "the ghost in the machine," is manifested by the brain, which it uses (via the bodily senses) to acquire and store information about the world and to integrate this with its genetically-driven imperative to live, thus resolving internal conflicts among action-choices and expressing these in what we (in our consciousness) experience as decisive action.

Scientific medicine has concentrated primarily on the body. The ancient Roman physician Galen first dissected and vivisected a variety of animals to increase his knowledge of anatomy and physiology, and dissection became increasingly important in the training of physicians and surgeons, and in painting and sculpture, during the Renaissance. By the late 20th century, dissection had reached the molecular level, with the insides of the human cell and nucleus being taken apart and examined by molecular biologists, literally receptor by receptor. Dissection and the mechanistic understanding it provides have led some to decry what they regard as the modern "soulless" view of the human body as a mere machine.

In the nanomedical era, even the most diehard reductionist must come to see the human body not merely as a heap of parts but rather as a finely tuned vehicle that is owned and piloted by a single human mind. As with automobiles, some body-owners will be more diligent about maintenance, regular tuneups, and paint jobs than other body-owners. Some will crave the latest upgrades, while others may prefer a more conservative model that reliably gets them around town. At either extreme, all human vices and virtues will be on full display, though one may perhaps anticipate an increasing pride of corporeal ownership if for no other reason than because maintenance and repair will become quick, convenient, and inexpensive.

From this simple analogy of body-and-mind to car-and-driver, it might at first appear that the advent of nanomedical technology will confirm and strengthen the traditional dualist conception of the body. But closer inspection reveals that the analogy is at best incomplete, and at worst deeply flawed. This is because mind, first being necessarily embedded in physical structure and relying upon that structure for its faithful execution, and second, this physical structure now being manipulable at the molecular level, enters also into the purview of our mechanic. Both car and driver may be modified in the shop. Speaking allegorically, it is as if the driver, after getting his car a tuneup, emerges from the shop no longer favouring chocolate but enjoying vanilla instead, or now preferring jazz over classical, the opposite of before. Such psychological changes may be either volitional or emergent.

Until the late 20th century, human progress was measured almost exclusively in terms of externalities. Food was gathered, then sown, then manufactured. Shelters had no running water, then gained outhouses, then indoor plumbing. Natural lighting and campfires gave way to candles, then oil lamps, then electric illumination. Finger-counting yielded first to the abacus, then the mechanical adding machine, and finally to the digital computer. But throughout all of history, the human body itself has remained largely untouched by progress. We have always regarded our bodies, evolved by natural selection, as fundamentally inviolate and immutable—subject perhaps to various natural or traumatic degradations, but rarely to any significant intrinsic improvement on the timescale of human civilization.

Now we are set to embark upon an era in which our natural physiological equipment may for the first time in history become capable of being altered, improved, augmented, or rendered more comfortable or convenient, due to advances in medical technology. The physical human body may be one of the last bastions of "naturalness". It will also be one of the last elements in our common worldview to be modernized.

Our subjective experience of reality will shift by subtle degrees. For instance, all objective information about our physical surroundings has traditionally arrived in the conscious mind via the various natural senses such as hearing, sight, and smell. In the nanomedical era, machine-mediated sensory modalities may permit direct perception of physical phenomena well removed from our bodies in both time and space, or which are qualitatively or quantitatively inaccessible to our original natural senses. Perception will gradually expand to incorporate nonphysical phenomena including abstract models of mental software, purely artificial constructs of simulated or enhanced realities, and even the mental states of others. Such new perceptions will inevitably alter the way our minds process information.

But the winds of change will sweep deeper still, into our very souls. Like ants oblivious to the collective purpose of their colony, the billions of neurons in the human brain are all busily buzzing, wholly ignorant of the emergent plan. This is the physical, mechanical world of our electrochemical hardware. People also have thoughts, feelings, emotions, and volitions, a higher level in the data processing hierarchy which in turn is equally oblivious of the brain cells. We can happily think while being totally unaware of any help from our neurons. But nanomedicine will give us unprecedented systemic multilevel access to our internal physical and mental states, including real-time operating parameters of our own organs, tissues, and cells, and, if desired, the activities of small groups of (or even individual) neurons. Diverse parts of our selves previously closed to our attention may slowly conjoin and enter our conscious awareness.

Will this access promote an integrated identity or lead to hopeless confusion, or worse? Marvin Minsky, in his collection of essays *The Society of Mind*, persuasively argues that our selves or identities are in fact networks of semi-autonomous neurological "agencies" which sometimes cooperate and sometimes compete with one another. We think of ourselves as singular "persons," but we also experience "conflicting desires" and "differing viewpoints" within our minds that are, in Minsky's view, a direct experience of the multiplicity of our brain's neurostructures. Other models of the human mind suggest that our internal mental states, prospectively transparent via nanomedical augmentation, are diverse and intricate; Julian Jaynes is one of many writers who have drawn attention to profound dichotomies between the two cerebral hemispheres. The component-oriented personality models of Freud (e.g. ego/id/superego), Jung (e.g. archetypes), and Rank (e.g.

will/counterwill), and the identification of 4541 distinct personality traits by Allport and Odbert warn us that full access to our brain's architecture could be perilous.

More seriously, most of us suppose that we are endowed with free will. But if choices by free will are simply the resolution of conflicts of neurological subsystems, and we become consciously aware of those subsystems and are able to intervene in their processes, do we run the risk of runaway instabilities at the deepest levels of what we presently call our "minds"? Will we find that these instabilities are profound counterparts to the maladies we currently designate as epilepsy, or psychosomatic illnesses? In any redesigns of our brains which would involve opening doors to, quite literally, the ultrastructure of our thoughts, we could become "naked to ourselves" in ways that we can only vaguely speculate about at present. Along with any other dangers we might encounter, this will raise entirely new issues of the proper role of psychotherapy and the sanctity of personal privacy.

Repairs to the brain may be carefully monitored to ensure quality control and to verify intended results, as already proposed in another context. Major modifications might be strictly regulated, both to prevent abuse by unscrupulous third parties and also to forestall accidental or volitional alterations that could render the patient a significant threat to society. Nanomedical alterations to the brain and other physical systems may give us vastly expanded freedom to be who we choose to be, along with increased responsibility to make wise and informed choices. The ethical and legal aspects of these questions, as well as the scientific and psychological ones, are extremely important and should be thoroughly debated in the years and decades that lie ahead.

Nanomedical Perspective

Nanomedicine and Molecular Nanotechnology

A mature nanomedicine will require the ability to build structures and devices to atomic precision, hence molecular nanotechnology and molecular manufacturing are key enabling technologies for nanomedicine. The prefix "nano-" (from the Greek root nanos, or dwarf) means one-billionth (10^{-9}) of something. The term "nanotechnology" refers most generally to technology on the scale of a billionth of a meter, or a nanometer (a nanometer is ~6 carbon atoms wide). Similarly, the words "nanomachine," "nanorobot," "nanomotor" and "nanocomputer" may refer to complex engineered objects fabricated by positioning matter with molecular control.

Molecular engineering was discussed as an extension of bulk technologies in the 1960s and 1970s by von Hippel, von Foester, and Zingsheim. The phrase "Nano-Technology" was first used in print in 1974 by N. Taniguchi to refer to the increasingly precise machining and finishing of materials, progressing from larger to smaller scales and ultimately to nanoscale tolerances, following in the path of Feynman's proposed "top-down" approach, a schemata which persisted in Taniguchi's thinking throughout the 1980s and 1990s. In 1981, K.E. Drexler described a new "bottom-up" approach involving molecular manipulation and molecular engineering in the context of building molecular machines and molecular devices with atomic precision, a fundamentally different mind-set. Drexler again described molecular technology in 1982 and molecular mechanical devices in 1983, first using the word "nanotechnology" in 1985 and 1986 as synonymous with molecular technology, finally settling upon "molecular nanotechnology" in 1991 and "molecular machine systems" in 1992 to clarify that his concept involved working devices constructed with atomic precision, as distinguished from nanostructured bulk materials, micromachinery, polymeric self-assembly, pure biotechnology, nanolithography, Langmuir-Blodgett thin films, and the like. Drexler's definition—molecular nanotechnology as the three-dimensional positional control of atomic and molecular structure to create materials and devices with molecular precision—is the usage. The first known use of the term "nanomedicine" was in 1991 by Drexler, Peterson, and Pergamit in their popular book *Unbounding the Future*.

Is molecular nanotechnology possible? This question is explicitly addressed, but the bottom line is that molecular nanotechnology violates no physical laws and there exist many possible technical paths leading to useful results. Even by 1985, for example, G. Yamamoto had reported on molecular gears, describing "compounds that exist in conformations which are regarded as static meshed gears with two-toothed and three-toothed wheels and some of them behave as dynamic gears." H. Iwamura prepared a system that formed a chain of beveled molecular gears with ~GHz rotation rates. In 1998, it was generally accepted that molecular nanotechnology would be developed, although there was still some disagreement about how long it would take.

It is often noted that molecular biological systems are themselves nanomachines, constituting an existence proof for molecular nanotechnology. Indeed, biotechnology is one possible implementation pathway for molecular nanotechnology that is being pursued. Such comparisons have a long history. For example, Marcello Malpighi (1628-1694), a professor of medicine in Pisa who discovered the fine structure of the lungs and the capillaries using the microscope, once observed that "our bodies are composed of strings, thread, beams, levers, cloth, flowing fluids, cisterns, ducts, filters, sieves, and other similar mechanisms."

Parallels to living systems as molecular machines were drawn by Changeau, McClaire, Laing, Drexler, and Mitchell, inspiring early thinking in molecular manufacturing. For example, in 1991 Drexler observed:

"Technology-as-we-know-it is a product of industry, of manufacturing and chemical engineering. Industry-as-we-know-it takes things from nature — ore from mountains, trees from forests — and coerces them into forms that someone considers useful. Trees become lumber, then houses. Mountains become rubble, then molten iron, then steel, then cars. Sand becomes a purified gas, then silicon, then chips. And so it goes. Each process is crude, based on cutting, stirring, baking, spraying, etching, grinding, and the like."

"Trees, though, are not crude. To make wood and leaves, they neither cut, stir, bake, spray, etch, nor grind. Instead, they gather solar energy using molecular electronic devices, the photosynthetic reaction centers of chloroplasts. They use that energy to drive molecular machines—active devices with moving parts of precise, molecular structure—which process carbon dioxide and water into oxygen and molecular building blocks. They use other molecular machines to join these molecular building blocks to form roots, trunks, branches, twigs, solar collectors, and more molecular machinery. Every tree makes leaves, and each leaf is more sophisticated than a spacecraft, more finely patterned than the latest chip from Silicon Valley. They do all this without noise, heat, toxic fumes, or human labour, and they consume pollutants as they go. Viewed this way, trees are high technology. Chips and rockets are not."

Contemplating applications of nanotechnology to medicine, Brian Wowk concluded that 20th century physicians were in a predicament similar to that which would be faced by 18th-century engineers trying to maintain a 20th-century automobile—repairs would be crude at best, and breakdowns inevitable:

"Like primitive engineers faced with advanced technology, medicine must 'catch up' with the technology level of the human body before it can become really effective. What is the technology level? Since the human body is basically an extremely complex system of interacting molecules (i.e., a molecular machine), the technology required to truly understand and repair the body is molecular machine technology—nanotechnology. A natural consequence of [our achieving] this level of technology will be the ability to analyze and repair the human body as completely and effectively as we can repair any conventional machine today."

In *Engines of Creation*, Drexler drew inspiration from the cell's eye view to recognize that nanotechnology could bring a fundamental breakthrough in medicine. Noting that 20th century physicians relied chiefly on surgery and drugs to treat illness, Drexler explained:

"Surgeons have advanced from stitching wounds and amputating limbs to repairing hearts and reattaching limbs. Using microscopes and fine tools, they join delicate blood vessels and nerves. Yet even the best microsurgeon cannot cut and stitch finer tissue structures. Modern scalpels and sutures are simply too coarse for repairing capillaries, cells, and molecules. Consider 'delicate' surgery from a cell's perspective. A huge blade sweeps down, chopping blindly past and through the molecular machinery of a crowd of cells, slaughtering thousands. Later, a great obelisk plunges through the divided crowd, dragging a cable as wide as a freight train behind it to rope the crowd together again. From a cell's perspective, even the most delicate surgery, performed with exquisite knives and great skill, is still a butcher job. Only the ability of cells to abandon their dead, regroup, and multiply makes healing possible."

"Drug therapy, unlike surgery, deals with the finest structures in cells. Drug molecules are simple molecular devices. Many affect specific molecules in cells. Morphine molecules, for example, bind to certain receptor molecules in brain cells, affecting the neural impulses that signal pain. Insulin, beta blockers, and other drugs fit other receptors. But drug molecules work without direction. Once dumped into the body, they tumble and bump around in solution haphazardly until they bump a target molecule, fit, and stick, affecting its function. Drug molecules affect tissues at the molecular level, but they are too simple to sense, plan, and act. Molecular machines directed by nanocomputers will offer physicians another choice. They will combine sensors, programs, and molecular tools to form systems able to examine and repair the ultimate components of individual cells. They will bring surgical control to the molecular domain." By the end of the 20th century, mainstream military (DoD), NIH, NSF, and other international groups had begun to seriously consider the potential future applications of molecular nanotechnology in medicine. For example, in 1997 a panel of U.S. Department of Defense health science experts known as Military Health Service Systems (MHSS) 2020 concluded in its final report:

"If a breakthrough to a [molecular] assembler occurs within ten to fifteen years, an entirely new field of nanomedicine will emerge by 2020. Initial applications will be focused outside the body in areas such as diagnostics and pharmaceutical manufacturing. The most powerful uses would eventually be within the body. Possible applications include programmable immune machines that travel through the bloodstream, supplementing the natural immune system; cell herding machines to stimulate rapid healing and tissue reconstruction; and cell repair machines to perform genetic surgery."

The present book takes as its starting point the assumption that the mass production of nanomachines at modest cost is technically feasible, and then explores the medical implications of this assumption. Proposed systems presented in this trilogy are intended not as final engineering blueprints but merely as points of departure for further analysis and refinement. All designs and projected capabilities are, for the most part, conservatively drawn with generous safety margins, leaving a considerable volume of design space yet to be explored by more intrepid future investigators.

Nanomedicine: History of the Idea

Conclusive proof of the existence of atoms was not obtained until the close of the 19th century. This may explain why the idea of nanomedicine is an exclusively 20th century phenomenon. The first hint of it may be found in a famous 1929 essay written by J.D. Bernal:

"The discoveries of the twentieth century, particularly the micro-mechanics of the Quantum Theory which touch on the nature of matter itself, are far more fundamental and must in time produce far more important results. The first step will be the development of new materials and new processes in which physics, chemistry and mechanics will be inextricably fused. The stage should soon be reached when materials can be produced which are not merely modifications of what nature has given us in the way of stones, metals, woods and fibers, but are made to specifications of a molecular architecture. Already we know all the varieties of atoms; we are beginning to know the forces that bind them together; soon we shall be doing this in a way to suit our own purposes. The result—not so very distant—will probably be the passing of the age of metals and all that it implies—mines, furnaces, and engines of massive construction. Instead we should have a world of fabric materials, light and elastic, strong only for the purposes for which they are being used, a world which will imitate the balanced perfection of a living body."

Development of the concept of nanomedicine has followed two principal paths which Richard Smalley has termed "wet nanotechnology" in the biological tradition, and "dry nanotechnology" in the mechanical tradition. Both approaches were presaged in speculative fiction. The following abbreviated history focuses on nanomedicine largely to the exclusion of broader issues in molecular engineering, manufacturing and nanoscopy, and includes a number of inspirational, speculative, or fictional early references from non-refereed sources.

Biological tradition

The general idea of biological engineering stretches back at least to the mid-19th century, but the first science fiction story involving actual genetic engineering was "Proteus Island" (1936), written by the chemical engineer Stanley Weinbaum. Artificial engineered organisms subsequently appeared in minor roles in several stories, a notable example being the familiars employed by the fake witches in Fritz Leiber's "Gather, Darkness!" (1943), and A.E. van Vogt used "gene transformation" to create the superman in "Slan" (1940). The first artificially evolved creatures appeared in Theodore Sturgeon's "Microcosmic God" (1941), wherein a biochemist established conditions allowing accelerated artificial evolution, creating the Neoterics, a submillimeter-sized race of intelligent hypermetabolic creatures which could accomplish tasks very rapidly. The first artificially designed microcreatures appeared in James Blish's "Surface Tension" (1952). In this story, crash-landed dying human astronauts create a completely new form of humanity—tiny men and women reduced to protozoan size—who are seeded in the pools and puddles at the surface of the new planet, and who go on to master the biotechnology necessary to travel from one water puddle to another.

The scientific tradition of biological nanomachines for medical purposes began in 1964 when Robert Ettinger, an early cryonics pioneer, suggested that cellular-level or even molecular-level repair might be developed for life extension. Ettinger speculated that "...surgeon machines, working 24 hours a day for decades or even centuries, will tenderly restore the frozen brains, cell by cell, or even molecule by molecule in critical areas."

In 1965, the synthesis of artificial life was publicly proposed as a national goal by the president of the American Chemical Society, Professor Charles Price, who pointed out that many new types of life might be made, not "mere imitations" of biology as we know it.

In 1967, Isaac Asimov suggested the future possibility of "factories...where the working machinery consists of submicroscopic nucleic acids" and that a "repertoire of hundreds or thousands of complex enzymes" could be used to "bring about chemical reactions more conveniently than any methods now used" and also for "helping to construct life."

In 1968, G.R. Taylor cited the possibilities for genetic engineering and genetic surgery: "The microsurgery of DNA may possibly be achieved by physical methods: fine beams of radiation (probably laser light or pulsed X-rays) may be used to slice through the DNA molecule at desired points." He also cited predictions that bacteria would soon be programmed.

In 1969, J. White suggested that a modified virus could be used as a cell repair machine: "It has been proposed that appropriate genetic information be introduced by means of artificially constructed virus particles into a congenitally defective cell for remedy; similar means may be used for the more general case of repair. The repair program must use means such as protein synthesis and metabolic pathways to diagnose and repair any damage... [Information] can be preserved by specifying that the repair program incorporate appropriate RNA tapes into itself..."

In 1970, Jeon et al carried out "the reassembly of *Amoeba proteus* from its major components: namely nucleus, cytoplasm, and cell membrane," taken from three different cells.

In 1972, Ettinger proposed using genetic engineering to make microscopic biorobots: "Genetic engineering's most sensational impact will concern the modification of humans, but it will have other uses as well. Some of the "robots" that will serve us will need to be nanominiaturized." Existing organisms could be modified to make biologically-based programmable biorobots for medical applications: "If we can design sufficiently complex behaviour patterns into microscopically small organisms, there are obvious and endless possibilities, some of the most important in the medical area. Perhaps we can carry guardian and scavenger organisms in the blood, superior to the leukocytes and other agents of our human heritage, that will efficiently hunt down and clean out a wide variety of hostile or damaging invaders." Computerized cellular repair machines "must use means such as protein synthesis and metabolic pathways to diagnose and repair any damage...[Information] can be preserved by specifying that the repair program incorporate appropriate RNA tapes into itself...." Also in 1972, Danielli described various possibilities for generating new life forms via "life-synthesis" and genetic engineering, noting that "macromolecular engineering" might enable the development of very powerful and compact macromolecular computer systems.

In 1974, Halacy noted "some rather inglorious ways" to use "the miracle of artificial life," including potential capabilities for growing diverse items ranging from computers to airplanes. Morowitz suggested cooling cells to cryogenic temperatures in order to analyze and determine their structure. Artificial cells could also be assembled at such temperatures and then be set in motion by thawing. Morowitz further reported that microsurgery experiments on amoebas "have been most dramatic. Cell fractions from four different animals can be injected into the eviscerated ghost of a fifth amoeba, and a living functioning organism results."

In 1975, Richard Laing described the theoretical possibility of molecular machine self-replicators using molecular (data) tapes based on the idea of universal Turing machines, examining several ways that such "artificial organisms" might replicate themselves as "a vehicle for the exploration of broad biological possibilities."

In 1976, Donaldson presented the first detailed (and quite ambitious) list of biotechnological techniques that appeared necessary to achieve cell repair and might prove feasible, writing at a time that predated many current capabilities such as automated protein/DNA sequencing and synthesis, and most knowledge of restrictions on cellular developmental pathways and genetic programs and networks. For repair at the level of the cell, Donaldson's techniques would have included:

1. The ability to design enzymes to produce specific repair functions such as renaturing denatured proteins, joining broken lipoprotein complexes, annealing broken strands of DNA or RNA, reading proteins of existing or special types onto RNA and replicating them, and giving a cell the ability to metabolize new substrates, use novel cofactors, or construct essential amino acids;
2. Specially constructed bacteria or macrophages able to replicate themselves, spread throughout a specific target tissue, and carry out specific repairs according to the programs designed into their DNA/RNA; these could be designed to operate at unnatural

temperatures or to utilize metabolic pathways not presently found in nature;

3. The abilities to re-introduce lost DNA or lost organelles such as mitochondria into a cell, to introduce entirely new forms of organelles perhaps to perform specific repair functions, and to introduce new metabolic capacities into a cell;
4. The ability to modify at will the developmental program of a cell, as for instance to induce postmitotic cells such as neurons to divide, according to a specific program, forming daughter cells with specific properties; and
5. Several different types of repair bacteria able to work together in an integrated fashion, and linked together by chemical means (e.g. hormones), so as to apply optimal repairs to every body cell in order to (A) diagnose the precise nature of the damage, call other repair bacteria to its location, and report to the attending doctor that new types of repair bacteria other than those already introduced are needed, and (B) identify structures which must be preserved (e.g. memory) and reconstruct them if necessary.

For repairs at the level of the whole organism, Donaldson offered the following rather aggressive biological nanotechnology techniques:

1. Understanding the physiology of aging combined with the ability to reverse it;
2. Control over growth and development, including the ability to program types of growth and development which do not naturally occur, such as growth of new eyes or other organs which have been lost or damaged, growth of an entire and well-formed body from a head alone, and regrowth of injured or lost brain tissue; and
3. Nonpermanent "substitute organs" which will take over from others which have been lost, including (A) the ability to keep a given tissue alive and healthy in vitro for an indefinite time and similar abilities for a body part, and (B) temporary replacements for any body organ which may have been lost. These would be used to support the body while new organs were growing, e.g. as "metabolic crutches." This capacity specifically includes the ability to make temporary replacements for diffuse systems such as the vascular or nervous systems. For instance, a specially created "plant" would grow an entire vascular system into the patient, down to replacement for the capillaries and venules from a single seed, always introducing its fibrils between cells and destroying none or very little of the original structure.

The last of these is a description of a (clearly speculative) "repair net," a concept which Donaldson may have been the first to propose in 1976. A whole-body repair net was later termed a "chrysalis", and twelve years later, in 1988, Donaldson provided an artist's conception and an additional description of these proposed biotechnological instrumentalities:

"Severe crushing or mangling injuries require us to provide a new vascular system. The repair device might resemble a fungus, growing mycelia into the injured tissue. A repair net would grow into a [crushed] limb, guided by recognition of the injured cells and a plan for how the limb should look after repair....[In the most severe injuries,] a chrysalis first envelops the patient, then enters in between all his cells. It disassembles the patient, surrounding each cell with its own repair machinery and vascular system. The geometry already preserves information about locations of the patient's cells. If necessary, morphogen chemical gradients could also retain this information. A patient would [locally] swell up to 10 times original diameter. After repair, the chrysalis withdraws the same way it entered."

In 1977, Darwin further developed the theme of tissue repair and cellular repair biomachines, independently proposing a modified white blood cell to perform repair functions.

In 1981, Asimov suggested that we "consider the bacteria. These are tiny living things made up of single cells far smaller than the cells in plants and animals....[We] can, by properly designing these tiniest slaves of ours, use them to reshape the world itself and build it close to our hearts' desire." More importantly, Donaldson elaborated on his earlier discussion of how cryonically suspended human beings might be repaired. He extended the earlier concepts of Ettinger and White, concluding that "with such hybrid technology as micro-miniature biological-mechanical machines the size of viruses" and related technology, "it seems unlikely that (to repair a single cell) there would be any difficulty at all in principle to carrying out any imaginable repair." He estimated that about 10 programmable cell-repair biomachines could be introduced into a cell that was under repair without causing too much mechanical disruption. In 1988, Donaldson described artificial macrophages that "can carry control machinery to recognize target cells, responding only to them, or responding differently depending upon cell type or cell conditions. They can still work even if the target cell isn't functioning (unlike viruses), rebuild target cell machinery other than the genes, and transfer many more genes up to an entire copy of the patient's genome." They can also "communicate with one another [and] release diffusible chemicals to guide one another's behaviour."

By the 1990s, bioengineered viruses of various types and certain other vectors were routinely being used in experimental genetic therapies as a means to target and penetrate certain cell populations, with the objective of inserting therapeutic DNA sequences into the nucleus of human target cells *in vivo*. Retrovirally-altered lymphocytes (T cells) began to be injected into humans for therapeutic purposes. Another example of an engineered cell in therapeutics was the use of genetically modified cerebral endothelial cell vectors to attack glioblastoma, which was being pursued by Neurotech (in Paris) in 1998. Engineered bacteria were being pursued by Vion Pharmaceuticals in collaboration with Yale University. In their "Tumour Amplified Protein Expression Therapy" program, antibiotic-sensitive *Salmonella typhimurium* (food poisoning) bacteria were attenuated by removing the genes that produce purines vital to bacterial growth. The tamed strain (cell line VNP20009) could not long survive in healthy tissue, but quickly multiplied ~1000-fold inside tumours, which are rich in purines. The next step would be to add genes to the bacterium to produce anticancer proteins that can shrink tumours, or to modify the bacteria to deliver various enzymes, genes, or prodrugs for tumour cell growth regulation. The engineered bacteria were available in multiple serotypes to avoid potential immune response in the host. Phase I human clinical trials were expected to begin in 1999 using clinical dosages produced in 50-liter fermenters, and other possible bacterial vector species were being examined.

Micro-biorobotics was still regarded by many in the biotechnology community as a highly speculative topic in 1998. Glen A. Evans described the possible construction of synthetic genomes and artificial organisms. His proposed strategy involves determining or designing the DNA sequence for the genome, synthesizing and assembling the genome, introducing the synthetic DNA into an enucleated pluripotent host cell, then introducing the host cell into an organism. Evans foresees the high throughput "read-out" of gene products discovered through genomic sequencing, rapid construction of designer genes and genomes, automated designer vector synthesis for gene replacement/insertional mutagenesis, and finally the construction of synthetic genomes and artificial organisms. Robert Bradbury has considered requirements and costs for genome synthesis and replacement. For example, Bradbury estimates genome synthesis costs by assuming a projected ~$0.03/

base for DNA synthesis (compared to ~\$0.80-\$1.00/base-micromole in 1999), with ~20,000 expressed genes per biorobotic cell with an average gene size of ~3000 bases, giving a ~60 megabase expressed genome per biorobotic cell (1 chromosome) and a raw synthesis cost of ~\$3 million per designer cell line (excluding design costs). There has also been discussion of "chemical reaction automata" as precursors for synthetic organisms, synthetic lifeforms and "nanobiology", "biorobotics", "cell rovers", microbial engineering, lymphocyte engineering, and artificial chromosomes, and cell engineering is rapidly gaining popularity.

Mechanical tradition

The late science fiction author Robert A. Heinlein nearly invented the concept of molecular nanotechnology in 1942 when he suggested a process for manipulating microscopic structures. Heinlein envisioned the extensive use of life-size teleoperator hands, called "waldoes," complete with sensory feedback for full, remote-controlled telepresence. His fictional hero, Waldo, used a collection of these mechanical teleoperated hands for building and operating a series of ever-smaller sets of such mechanical hands. The smallest mechanical hands, "hardly an eighth of an inch across," were equipped with micro-surgical instruments and stereo "scanners," and were used to "manipulate living nerve tissue, [to examine] its performance *in situ,*" and to perform neurosurgery. Eric Frank Russell's 1947 story "Hobbyist" described a fabrication process with "atom fed to atom like brick after brick to build a house." In 1955, Russell's serial "Call Him Dead," also published in *Astounding Science Fiction*, had a virus-based alien intelligence that spread through contact with blood or saliva; the story features a "microforger," a man who makes "surgical and manipulatory instruments so tiny they can be used to operate on a bacillus." Also in the mechanical tradition, Isaac Asimov's "Fantastic Voyage" in 1966 took its miniaturized human crew in a miniaturized submarine through the bloodstream of a human patient on a mission of repair.

Heinlein, Russell, and Asimov overlooked the full implications of their ideas, and most scientists were unsympathetic--for example, in 1952 Erwin Schrodinger wrote that we would never experiment with just one electron, atom, or molecule. But by the early 1960s, several scientists had reinvented similar approaches to micromanipulation and miniaturization, this time extending their reach into the nanotechnology domain. The first and most famous of these scientists was the Nobel physicist Richard P. Feynman. In his remarkably prescient 1959 talk

"There's Plenty of Room at the Bottom," Feynman proposed employing machine tools to make smaller machine tools, these to be used in turn to make still smaller machine tools, and so on all the way down to the atomic level. Feynman prophetically concluded that this is "a development which I think cannot be avoided." Such nanomachine tools, nanorobots and nanodevices could ultimately be used to develop a wide range of submicron instrumentation and manufacturing tools, i.e., nanotechnology. Feynman's suggested applications for these tools included producing vast quantities of ultrasmall computers and various micro- and nano-robots.

Feynman was clearly aware of the potential medical applications of the new technology he was proposing. After discussing his ideas with a colleague, Feynman offered the first known proposal for a nanomedical procedure to cure heart disease: "A friend of mine suggests a very interesting possibility for relatively small machines. He says that, although it is a very wild idea, it would be interesting in surgery if you could swallow the surgeon. You put the mechanical surgeon inside the blood vessel and it goes into the heart and looks around. (Of course the information has to be fed out.) It finds out which valve is the faulty one and takes a little knife and slices it out. Other small machines might be permanently incorporated in the body to assist some inadequately functioning organ." Later in his historic lecture, Feynman urged us to consider the possibility, in connection with biological cells, "that we can manufacture an object that maneuvers at that level!"

In 1961, K.R. Shoulders rejected the use of biological building blocks, even though biological "processes do work, and they can do so in a garbage can without supervision." He saw them as too limited environmentally and too difficult to control with available technology. Instead, he sought to directly produce much simpler, more powerful and rugged nanostructure arrays, operating at video frequency rates, which in turn could ultimately aid in their own replication. In 1965, Shoulders reported the actual operation of micromanipulators able to position tiny items with 10 nm accuracy while under direct observation by field ion microscopy.

In 1970, Volkenstein noted that "the creation of a nonmacromolecular system which would act as a model for living organisms is definitely possible" but could not arise by itself, and that the macromolecularity of present organisms is not essential, but due to their evolutionary origins. Taking some poetic license, he added: "Consequently, the cybernetic nonmacromolecular machine, which simulates life, could have been and can be created on earth only by man. Then it could perfect itself without limits." T. Nemes discussed artificial self-replicating machines and described how to construct "an automatic lathe able to reproduce itself," a concept apparently developed before von Neumann's work on machine replication.

In 1981, Drexler suggested the construction of mechanically deterministic nanodevices using biological parts; these devices could inspect cells at the molecular level and also repair cellular tissues that had been damaged during cryonic suspension. In 1982, Drexler described cell repair machines even more clearly in the mechanical tradition, in a popular publication.

By 1983, Drexler began privately circulating a draft technical paper entitled "Cell Repair Machines" which investigated for the first time, in some detail, whether an advanced mechanical-based nanotechnology would "permit construction of systems of molecular-scale sensors, computers, and manipulators able to enter and repair cells; the nature of the computational algorithms and magnitude of the computational resources needed to guide repairs; and the physical capabilities and constraints important to the repair process [in order to] sketch the conceptual design of a cell repair system based on a mature molecular technology."

Peterson notes that "medical applications were explored by Drexler during the early 1980s but the medical community was not ready for the concept." A technical paper by Drexler, invited by an editor at the *Journal of the American Medical Association*, was dismissed by a referee as "science fiction."

In 1985, G. Feinberg proposed using short-wavelength coherent laser energy to power and communicate with "nanosensors that could be implanted into the human body. They could...[monitor] various physiological functions from subcellular molecules up through tissues and organs...essential in determining some of the mechanisms involved in growth and aging."

In a 1985 book entitled *Robotics*, edited by Marvin Minsky (a well-known computer scientist and artificial intelligence pioneer), Minsky briefly described how fully-automatic cellular repair machines might work, following Drexler's vision:

"Suppose that we could design a repair machine so small that it could repair an artery from the inside! The first such machines might be the size of fleas. (There is room for a great deal of machinery in something the size of a flea—as the body of the flea itself testifies.)

These micromachines could crawl into all but the smallest blood vessels, clear out debris, and reline the walls with suitable materials yet to be invented. Later generations of micromachines could be even smaller, perhaps no larger than the body cells they repair. These minuscule machines would be mass-produced by the billions, either by larger machines or by techniques of making them reproduce themselves. Perhaps these biological janitors would even be implanted in our bodies to remain there as permanent maintenance workers, just like many biological cells that already serve such purposes."

"Today the idea of such a technology may seem fantastic, yet many of the circuits in our computers are already smaller than many of our bodies' cells. Let's try, for a moment, to look ahead to: mass production of highly intelligent machines; gigantic advances in miniaturization; a technology so advanced that these machines reproduce themselves without our help. Fantastic? Not at all. Even the simplest algae and bacteria can do that. True, they're not intelligent, but each of them contains enough computerlike machinery and memory to do those things. So, to build bacteria-size computers should be perfectly feasible, once we have the necessary microtechnology. Some day we'll have the means to build artificial, cell-like machines with all those capabilities."

R.A. Freitas Jr. also contributed a chapter to Minsky's 1985 book, offering the following suggestion for remote-controlled incisionless nanosurgery: "One possibility is the concept of remote-controlled medical mites made feasible by modern micromachinery technology. Some medical mites would be like microminiature submarines, released inside the human body for internal sensing. Other mites could float, crawl, or swim through major arteries in the human body and perform on-site repairs from within, controlled by radio link under direction of a skilled telemicrosurgeon."

In 1986, Drexler published *Engines of Creation*, a popular text with two chapters devoted to discussions of cellular repair machines. Drexler further expounded upon the topic of cellular repair machines in articles published in 1985, 1986, 1987, and 1989, and his concepts were described by Brian Wowk in 1988 and in a Time-Life book in 1989.

In 1988, A.K. Dewdney reported an early nanomedical concept of an artery-cleaning nanorobot that he attributed to Drexler, accompanied by an artist's conception with the caption "a nanomachine swimming through a capillary attacks a fat deposit."

Nanomedicine in the 1990s

Despite the tremendous importance of molecular nanotechnology, a published 1993 literature review of robotics in health care included not a single reference to nanotechnology or to nanomedicine. At this writing in 1998, the list of original post-1989 technical and popular nonfiction works that deal with nanomedical topics in the mechanical tradition is short enough to permit virtually an exhaustive listing, including Beardsley, Bova, Coombs and Robinson, Crawford, Drexler, DuCharme, Emanuelson, Fahy, Fiedler and Reynolds, Freitas, Kaehler, Klatz and Kahn, Kurzweil, Lampton, Merkle, Merrill, Minsky, More, Ostman, Reifman, and Wowk and Darwin. The first nanomedical device design technical paper was published in 1998 by Freitas in the biotechnology journal *Artificial Cells*, and the present trilogy (*Nanomedicine*) is the first book-length technical treatment of the medical implications of molecular nanotechnology.

Biotechnology and Molecular Nanotechnology

Some pragmatic readers may be wondering what is so special about molecular nanotechnology, that existing or anticipated biotechnology could not accomplish just as well? After all, biotechnology is already an established medical capability. It has real applications and real products already on the market. Reflecting upon the future possibility of sophisticated mechanical medical nanorobots equipped with powerful nanocomputers, in 1989, one well-known cryobiologist mused that "what is not clear is just what need we would have for such devices." The simplest answer, is that each of the three contemporary branches of "nanotechnology" offers something of unique value to the practice of medicine.

Nanoscale materials technology has already found widespread use in medicine, including biocompatible materials and analytical techniques, surgical and dental practice, nerve cell research using intracellular electrodes, biostructures research and biomolecular research using near-field optical microscopy, scanning-probe microscopy and optical tweezers, and vaccine design, and also many 20th century bulk chemical and biochemical manufacturing techniques along with much of classical pharmacology.

As for "biotechnology," the original meaning of this word contemplated "the application of biological systems and organisms to technical and industrial processes". In recent times, the field has expanded to include genetic engineering and now takes as its ultimate goal no less than the engineering of all biological systems, even completely artificial organic living systems, using biological instrumentalities.

The third branch, molecular nanotechnology, takes as its purview the engineering of all complex mechanical systems constructed from the molecular level—potentially offering new tools for medical practice, the principal subject of this book. Observes G.M. Fahy, "the difference between nanotechnologists and biotechnologists is that the former do not restrict themselves to the biological limitations of the latter, and they are much more ambitious about the kinds of accomplishments that they want to achieve."

Doctors can utilize solutions to medical problems from all three approaches. As noted earlier, 80-90% of medical complaints resolve themselves via natural homeostatic processes, or without the necessary involvement of active biotechnological or molecular-nanotechnological agents. But by employing biotechnology, the range and efficacy of treatment options greatly increases. With molecular nanotechnology, the range, efficacy, comfort and speed of possible medical treatments again expands enormously. Molecular nanotechnology is essential when the damage to the human body is extremely subtle, highly selective, or time-critical (as in head traumas, burns, or fast-spreading diseases), or when the damage is very massive, overwhelming the body's natural defenses and repair mechanisms.

At every difficulty level, most classes of medical problems may be resolved with varying efficacy within the homeostatic/nanomaterials, biotechnological, or molecular-nanotechnological approaches. As the chosen technology becomes more precise, active, and controllable, the range of options broadens and the quality of the options improves. Thus the question is not whether molecular nanotechnology is required to accomplish a given medical objective. In many cases, it is not—though of course there are some things that only biotechnology and molecular nanotechnology can do, and some other things that only molecular nanotechnology can do. Rather, the important question is which approach offers a superior solution to a given medical problem, using any reasonable metric of treatment efficacy. For virtually every class of medical challenge, a mature molecular nanotechnology offers a wider and more effective range of treatment options than any other approach.

It is quite possible to imagine an advanced biotechnology that uses an engineered white cell, fibroblast, or macrophage chassis, energized by native oxygen and glucose and modified mitochondrial powerplants, driven by pseudopodia, cilia or flagella, communicating and navigating via biochemical signals, and even incorporating onboard digital biocomputers to make microscopic biorobots. Principal arguments favouring the biotechnology approach for medical purposes are: (1) that we are already somewhat familiar with such systems, after half a century of intensive molecular biology research; (2) that we have already "built" precursor systems, such as a whole living amoeba constructed from five distinct parts, bioengineered viruses and bacteria as DNA insertion devices, and natural replication stimulated in genetically engineered starter microbes; (3) that biocompatibility will not be a major issue since fibroblasts (which express no HLA Class II antigens, hence stimulate no rejection response) could be used as the starter material; (4) that both engineered viruses and bacteria are already in wide commercial and research use; and (5) the greater complexity of self-repair in mechanical systems, should it be needed. Many believe that the development pathway to early biorobots may be considerably shorter than for the mechanical nanorobots of the molecular nanotechnology approach, for which in 1998 not a single working prototype yet existed, even in research laboratories.

It is also possible to imagine a molecular nanotechnology that uses mechanical nanorobotic systems. Such systems will have many constitutional differences from biological-based systems. For instance, mechanical systems will transport parts, materials, energy and instructions via fixed channels, whereas most (but not all) biological systems operate by diffusion. Mechanical systems will have structures constrained by specific geometries, whereas biological systems have structures defined by patterns of containment and interconnection—the shape of a membrane compartment in a cell matters less than its continuity and the contents of the volume it defines.

Mechanical systems will be deterministically manufactured by operations analogous to manual construction, whereas engineered ribosomes will self-assemble via diffusion and stochastic matching of complementary parts; in other words, biology uses recipes, while mechanical systems use blueprints. Cells grow, with their parts adapting to one another; mechanical nanorobots may be constructed from parts of fixed structure. Biology uses self-repair; mechanical systems generally do not largely because self-repair (by component exchange) will not be needed in molecular mechanical systems, whose designs may be made more simple by relying upon high component redundancy. These and other differences imply a number of important advantages that mechanical-based medical systems will enjoy over biological-based medical systems, which, taken together, strongly suggest that predominantly mechanical nanosystems may be the approach of choice for a mature medical nanotechnology. The advantages of molecular nanotechnology (e.g. the mechanical tradition) are many.

Speed of medical treatment

Doctors may be surprised by the incredible quickness of nanorobotic action when compared to the speeds available from fibroblasts or leukocytes. Normal homeostatic processes such as dermal wound repair via natural fibroblasts may require weeks to run to completion. Typical fibroblast movements occur at 0.1-1 microns/sec, but mechanical nanomanipulators can operate at 1-10 cm/sec speeds or faster, a speed advantage of 4-5 orders of magnitude. Even the strongest biological fibers (e.g. intermediate filaments) have a failure strength 3 orders of magnitude below the strongest mechanical fibers (e.g. fullerene nanotubes). Biological cilia beat at ~30 Hz while mechanical nanocilia may cycle up to ~20 MHz, though practical power restrictions and other considerations may limit them to the ~10 KHz range for most of the time. Flexible mechanical surfaces can complete a morphing motion in ~0.1 millisec, compared to the ~100 millisec snapback time for pinched red cell membrane, again a thousand-fold advantage in speed. Thus we expect that mechanical therapeutic systems can reach their targets up to ~10,000 times faster, all else equal, and treatments which require ~10^5 sec for a biological system may need only ~10^2 sec for a mechanical system; tachyiatria improves both patient and physician comfort. In either biological or mechanical systems, large numbers of devices of comparable physical size (e.g. ~microns) may be employed to do the work, so numbers alone cannot offset the mechanical speed advantage.

Power density and transduction

Biological cells typically employ power densities of 10^3-10^4 W/m^3, with maximum densities of ~10^6 W/m^3 in honeybee flight muscle cells and bacterial flagellar motors. By contrast, nanomechanical power systems can produce power densities of 10^9-10^{12} W/m^3, an advantage of 10^3-10^8 for mechanical over biological systems. By 1998, conducting polymer-based actuators generated 20-100 times the force for a given cross-sectional area as mammalian skeletal muscle. Additionally, amoebic locomotion in motile cells requires diffusion-limited cytoskeletal disassembly and reassembly to achieve movement; mechanical motility systems may employ simple cable-pulling, winches, or ratchets, which are faster and more direct. Some biological energy transducers are reversible, but muscle contraction is irreversible—not only cannot muscle actively re-expand, but stretching it doesn't make it produce much useful chemical energy. In contrast, electric motors can be run backwards to generate electricity, forcing pistons makes them pump, and loudspeakers can be used as microphones.

Superior building materials

Typical biological materials have tensile failure strengths in the 10^6-10^7 N/m^2 range, with the strongest biological materials such as wet compact bone having a failure strength of $\sim 10^8$ N/m^2, all of which compare poorly to $\sim 10^9$ N/m^2 for good steel, $\sim 10^{10}$ N/m^2 for sapphire, and $\sim 10^{11}$ N/m^2 for diamond and carbon fullerenes, again showing a 10^3-10^5 advantage for mechanical systems that use nonbiological materials. Nonbiological materials can be much stiffer, permitting the application of higher forces with greater precision of movement, and they also tend to remain stable over a larger range of temperature, pressure, salinity and pH. Proteins are heat sensitive in part because much of the functionality of their structure is due to noncovalent bonds involved in folding, which are broken more easily at higher temperatures; in diamond, sapphire, and many other rigid materials, structural shape is covalently fixed, hence is far more temperature-stable. Most proteins tend to become dysfunctional at cryogenic temperatures, unlike diamond-based mechanical structures. Biomaterials are not ruled out for all nanomechanical systems, but represent only a small subset of the materials that can be used in nanorobots. Mechanical systems can employ a wider variety of atoms and molecular structures in their design and construction, with novel functional forms that might be difficult to implement in a biological system such as steam engines or nuclear power. As another example, an application requiring the most effective bulk thermal conduction possible should use diamond, the best conductor available, not some biomaterial with inferior thermal performance.

Nondegradation of treatment agents

Diagnostic and therapeutic agents constructed of biomaterials generally are biodegradable in vivo, although there is a major branch of pharmacology devoted to designing drugs that are moderately non-biodegradable—anti-sense DNA analogues with unusual backbone linkages and peptide nucleic acids (PNAs) are difficult to break down. However, suitably designed nanorobotic agents constructed of nonbiological materials are not biodegradable. An engineered fibroblast may not stimulate an immune response when transplanted into a foreign host, but its biomolecules are subject to chemical attack *in vivo* by free radicals, acids, and enzymes. Even "mirror" biomolecules or "Doppelganger proteins" comprised exclusively of unnatural D-amino acids have a lifetime of only ~5 days inside the human body. Nonbiological materials such as diamond and sapphire are highly resistant to chemical breakdown or leukocytic degradation *in vivo*, and pathogenic biological entities cannot easily evolve useful attack strategies against these materials.

Control of nanomedical treatment

Present-day biotechnological entities are not programmable and cannot be switched on and off conditionally during task execution. A digital biocomputer, while possible in theory, represents a considerable conceptual departure from the usual biological paradigm. Even assuming that a digital biocomputer could be installed in a fibroblast, and that appropriate effector mechanisms could be attached, such a system would necessarily have slower clock cycles, less capacious memory per unit volume, and longer data access times, implying less diversity of action, poorer control, and less complex executable programs than would be available in nanoscale electromechanical computer systems.

The mechanical approach emphasizes precise control of action, including control of physical placement, timing, strength, structure, and interactions with other (especially biological) entities. The biological approach emphasizes the use of poorly controlled natural structures, needlessly sacrificing huge blocks of the available functionality and design space.

Nanodevice versatility

Mechanical systems can readily incorporate biological elements if necessary, but artificial biological systems can incorporate nonbiological materials such as carbon nanotubes or diamond/sapphire structural elements only with difficulty, in part because biology has a more limited repertoire of "effector" mechanisms. Artificial biological systems cannot easily incorporate nonbiological materials where desired because natural biological assembly methods make no provisions for these materials either in the coded instructions in DNA or in the attachment chemistries.

Rebuilding or reconstructing the human body with nonbiological components (e.g. fullerene encabled bone for bone damaged in an accident), or augmentation of human body function with unnatural abilities or features (e.g. autogenous paracrine control), will be very difficult or impossible to achieve using purely biotechnological means.

Avoiding overspecialization

M. Krummenacker notes that one of the most glaring shortcomings of bacteria and other naturally occurring molecular machinery—when viewed as systems subject to further engineering—is the rather limited range of molecular substrates they can utilize. For example, bacterial enzymes are highly specialized devices, with very narrow substrate specificities. Thousands of different enzymes are needed in each organism, and the substance classes capable of digestion are limited to some sugars, various amino acids including proteins that can be degraded by excreted proteases, lipids, and a few other smallish oxygen-functionalized carbon molecules such as glycerol and ethanol. Some bacteria can metabolize CO_2 and a handful of aromatic compounds, but there is a vast range of organic chemicals that most bacteria cannot degrade or manufacture. A much smaller set of substantially more general molecular tools can probably be designed using the mechanical approach; mechanosynthesis can fabricate and assemble, or disassemble, a far wider range of molecular structures than are available to the cellular machinery of life.

Faster and more precise diagnosis

The analytic function of medical diagnosis requires rapid communication between the injected devices and the attending physician. If limited to chemical messaging, biotechnology devices will require minutes or hours to complete each diagnostic loop. Nanomachines, with their more diverse set of input-output mechanisms, can outmessage the results of *in vivo* reconnaissance or testing literally in seconds. Such nanomachines can also run more tests of greater variety in less time. Mechanical nanoinstruments, including molecule-by-molecule disassemblers, will make comprehensive cell mapping and cell interaction analysis possible. Bacterial resistance can be assayed at the molecular level, allowing new treatment agents to more easily be composed, manufactured and immediately deployed.

More sensitive response threshold for high-speed action

Unlike natural systems, an entire population of nanobiotic devices can be triggered globally by just a single local detection of the target antigen or pathogen. The natural immune system takes $>10^5$ sec to become fully engaged after exposure to a systemic pathogen or other antigen-presenting intruder. A biotechnologically enhanced immune system that can employ the fastest natural unit replication time ($\sim 10^3$ sec for some bacteria) will require $\sim 10^4$ sec for full deployment post-exposure. By contrast, a nanobiotic immune system can probably be fully engaged (though not finished) in at most two blood circulation times, or $\sim 10^2$ sec.

More reliable operation

Engineered macrophages would probably individually operate less reliably than mechanical nanorobots. Many pathogens, such as *Listeria monocytogenes* and *Trypanosoma cruzi*, are known to be able to escape from phagocytic vacuoles into the cytoplasm; while biotech drugs or cell manufactured proteins could be developed to prevent this (e.g. cold therapy drugs are entry-point blockers), nanorobotic trapping mechanisms can be more secure. Proteins assembled by natural ribosomes typically incorporate one error per $\sim 10^4$ amino acids placed; current gene and protein synthesizing machines utilizing biotechnological processes have similar error rates. A molecular nanotechnology approach will improve error rates by at least a millionfold. Mechanical systems can also incorporate sensors to determine if and when a particular task needs to be done, or when a task has been completed. Finally, it is unlikely that natural organisms will be able to infiltrate mechanical nanorobots or to co-opt their functions. By contrast, a biological-based robot could be diverted or defeated by microbes that can piggyback on its metabolism, interfere with its normal workings, or even incorporate the device wholesale into their own structures, causing the engineered biomachine to perform some new or different function than was originally intended. There are many examples of such co-option among natural biological systems, including the protozoan mixotrichs found in the termite gut that have assimilated bacteria into their bodies for use as motive engines, and the nudibranch mollusks (marine snails without shells) that steal nematocysts (stinging cells) away from coelenterates such as jellyfish (i.e. a Portuguese man-of-war) and incorporate the stingers as defensive armaments in their own skins, a process which S. Vogel has called "stealing loaded guns from the army."

Verification of progress and treatment

Using a variety of communication modalities, nanorobots can report back to the attending physician, with digital precision, a summary of diagnostically- or therapeutically-relevant data describing exactly what was found, and what was done, and what problems were encountered, in every cell visited. A biological-based approach relying upon chemical messaging is necessarily slow with limited signaling capacity. Also unlike mechanical nanorobots, biotechnological systems generally cannot monitor their own functions while working, and, except for a few highly specialized DNA proofreading systems, cannot directly inspect their work while it is in progress or after it is finished.

Minimum side effects

Almost all drugs have side effects, such as conventional cancer chemotherapy which causes hair loss and vomiting, although computer-designed drugs have high specificity and relatively few side effects. Carefully tailored cancer vaccines under development in the late 1990s were expected unavoidably to affect some healthy cells. Even well-targeted drugs are distributed to unintended tissues and organs in low concentrations, although some bacteria can target certain organs fairly reliably without being able to distinguish individual cells. By contrast, mechanical nanorobots may be targeted with virtually 100% accuracy to specific organs, tissues, or even individual cellular addresses within the human body. Such nanorobots should have few if any side effects, and will remain safe even in large dosages because their actions can be digitally self-regulated using rigorous control protocols that affirmatively prohibit device activation unless all necessary preconditions have been, and continuously remain, satisfied.

G.M. Fahy has noted that these possibilities could transform "drugs" into "programmable machines with a range of sensory, decision-making, and effector capabilities [that] might avoid side effects and allergic reactions... attaining almost complete specificity of action....Designed smart pharmaceuticals might activate themselves only when, where, and if needed." Additionally, nanorobots may be programmed to excuse themselves from the site of action, or even from the body, after a treatment is completed; by contrast, spent biorobotic elements containing ingested foreign materials may have more limited post-treatment mobility, thus lingering at the worksite causing inflammation when naturally degraded or removed.

Reduced replicator danger

Drexler points out that living systems are evolved systems, while nanomechanical replicators would be designed: "The former are shaped to serve the goal of their own survival and replication in a natural environment, whereas the latter will be shaped (whether well or poorly) to serve human goals, perhaps in an artificial environment." Genetic engineering involves not design of replicators from scratch, but tinkering with the molecular machinery of existing bioreplicators. Since bioreplicators were not designed, they are not necessarily structured in ways that lend themselves to complete understanding, and processes based on diffusion and matching allow complex nonlocal interactions that can be hard to trace. Bioreplicators can be crippled, but having evolved in nature, they resemble systems that can survive in nature. Typically, they are able to exchange genetic information with wild organisms, raising the possibility of the introduction of new, unconstrained replicators in the natural environment.

Having evolved to evolve, they have a capacity for further evolution—to serve their own survival, not human goals. For example, even when stripped of key pieces of DNA to interfere with its replication powers, a live attenuated AIDS vaccine can slowly recover its virulence and can attack immune cells. R. Bradbury suggests that artificial biorobots could incorporate multiple fail-safe mechanisms including required external essential nutrients or suicide suppressors, self-destruct triggers, countdown timers like telomeres, and engineering to reject foreign DNA (the basis for restriction enzymes), but it remains logically easier for a system with inchoate capacity to evolve to resume doing so, than for a system which has never had this capability to spontaneously develop it. In contrast, nanomechanical replicators (e.g. assemblers) will be designed from scratch and thus will differ fundamentally from biological systems. The parts and structures of designed mechanical systems will be known, and the relationships among their parts will also be designed and fixed.

More important, nanoreplicators will be fundamentally alien to the biosphere, unrelated to anything that has evolved to survive in nature. Certainly the capacity to fail can appear by accident, and emergent capabilities cannot be completely ruled out, but engineering experience shows that the ability to perform complex organized activities (such as replication in a natural environment) does not normally appear spontaneously. Also, and purely as a geometrical consideration, adding a new part inside a densely organized geometric structure (as would be found in a nanomachine) typically requires changes in the relative positions of many other parts, and hence corresponding adjustments in design. Adding a part inside a densely organized topological structure typically leaves topologies unchanged—room can be made by stretching and shifting other parts, with no change in their essential design, hence permitting easier modification to biological structures by exogenous agencies. Note that mechanical medical nanodevices need not be capable of replication. There is no requirement for replication *in vivo*; such replication would be needlessly dangerous, and adding this capability would reduce effectiveness in carrying out the primary medical task. Analogously, viral vectors employed in genetic therapies are modified to be "incapable" of replication.

Assured patentability

Microscopic biorobots, unavoidably derived from natural biological material, may someday be deemed unpatentable under a general prohibition on "genetic colonialism" or other emerging legal doctrines. In contrast, mechanical nanorobots, being fully-artificial and designed machines, should always be patentable provided they satisfy the customary legal criteria.

Naturophilia

As already noted, living things in general and the human body in particular are awesome examples of a powerful and intricately woven natural molecular technology to which human engineers, in 1998, still aspire. But embracing Nature is not the same as finding her perfect. Today, the word "natural" has acquired a strong connotation of rightness, even of sanctity. For most of human history, notes biologist Steven Vogel, "the natural and human worlds stood opposed. Nature was something to be tamed and utilized; we had the ordinary attitude of organisms toward other species. Nowadays the natural world intrudes far less but gets venerated far more. And why not? When one's meat is bought in a store, when locusts don't threaten one's corn crop, when central heating and plumbing are the norm, the aesthetics of nature hold greater appeal." And so we embrace the natural rectitude or moral superiority of nature's ways, a kind of pantheism which may be called ethical naturalism, biophilia, or naturophilia.

Many great minds have fallen prey to naturophilia. In the 4th century BC, Aristotle wrote that "if one way be better than another, that you may be sure is Nature's way." In the 15th century, we have from Leonardo da Vinci: "Human ingenuity may make various inventions, but it will never devise any inventions more beautiful, nor more simple, nor more to the purpose than Nature does; because in her inventions nothing is wanting and nothing is superfluous."

However, as Virginia Postrel notes in *The Future and Its Enemies*: "If nature is itself a dynamic process rather than a static end, then there is no single form of 'the natural.' An evolving, open-ended nature may impose practical constraints, but it cannot dictate eternal standards. It cannot determine what is good. The distinction between the artificial and the natural must lie not in their source—human or not—but in their characteristics, in the way they relate to the world around them."

According to the dictionary, "artificial" usually means "made by man, rather than occurring in nature." More usefully, Herbert Simon defines the artificial as that which is designed, expresses goals, and possesses external purposes. The artificial is controlled and serves its creators' purposes, subject to the universal laws of physics. Kevin Kelly defines the natural as "out of control." Nature is evolved, not designed, and serves no goal or external purpose save its own survival. Nature, lacking intent, is amoral—it simply is. By building the artificial, observes Postrel, "we do not overthrow nature, but cooperate with it, using nature's own art to create new natural forms. Our artifice alters the path of nature, but it does not end it, for nature has no stopping point, no final shape. It is a process, not an end."

Some naturophilist writers have decried the increasing "medicalization of society" in which formerly natural functions have come to be regarded as medical conditions requiring intervention or treatment. However, history suggests that naturophilia is usually undermined by any new medical technology that offers clear, safe, and immediate benefits to patients. For example, prior to 1842, intense pain was viewed as the natural outcome of being cut with a scalpel during surgery. It had always been so—how could it ever be otherwise? The invention of anesthesia in 1842 suddenly altered this natural outcome and replaced it with a less painful artificial outcome, despite anguished cries from naturophiles within the medical community that eliminating pain might somehow diminish the human character.

Another example of a widely-accepted medicalization of normal function is childbirth, a quite natural activity that can nevertheless be very dangerous to the mother's health. Precise prehistoric death rates are unknown, although archeological evidence shows that Neandertal females tended to die before the age of 30 due to hazards of childbirth. In the worst 19th century maternity hospitals the natural death rate from childbirth was 9-10%, falling to a very artificial 0.4% rate in England by 1930 and to less than 0.01% in the U.S. during the 1990s. As a result, it now seems "natural" for a woman always to survive childbirth, even though the reverse may have been true for most of human history. Warns historian Roy Porter: "We should certainly not hanker after some mythic golden age when women gave birth naturally, painlessly, and safely; the most appalling Western maternal death rate today is among the Faith Assembly religious sect in Indiana, who reject orthodox medicine and practice home births; their perinatal mortality is 92 times greater than in Indiana as a whole."

A disease seems "natural" to those who suffer from it when no treatment exists. But once a treatment is discovered and is widely employed, the disease becomes rare and its absence now becomes "natural." To those in the past, writes K.E. Drexler, "the idea of cutting people open with knives painlessly would have seemed miraculous, but surgical anesthesia is now routine. Likewise with bacterial infections and antibiotics, with the eradication of smallpox, and the vaccine for polio: each tamed a deadly terror, and each is now half-forgotten history. What amazes one generation seems obvious and even boring to the next. The first baby born after each breakthrough grows up wondering what all the excitement was about." In the next century, says Charles Sheffield, "our descendants will look on angiograms, upper and lower GIs, and biopsies the way we regard the prospect of surgery without anesthetics."

Future generations who take for granted an all-pervasive nanomedicine in their lives may look aghast upon the 20th century, wondering among other things how we managed to retain productive focus given the constant annoyance of our numerous undiagnosed minor disease states. Most of these diseases are not yet recognized as such, and many are still regarded as "natural" and not worthy of treatment. In a few decades, this may change. Some examples:

Addictions

In 1998, many people laugh off seemingly harmless addictions to chocolate (*chocoholics*), fats or sugar (*sweet tooth*), food (*gluttony*), nicotine (*smokers*), caffeine (*coffee* and *cola drinkers*), work (*workaholics*), exercise (*runner's high*), telling falsehoods (*pathological liars*), gambling (*wagerphilia*), stealing (*kleptomania*), medical treatments (*hypochondria*), marriage (*polygamy*), power-seeking (*domination*), skydiving or bungee-jumping (*thrillseeking*), superstition (*astrology*), shopping (*spendoholics*), driving cars that kill 40,000 Americans per year (*mobilophilia*), unusual sexual preferences (bestiality), sexual activity (*nymphomania*, *satyriasis*), or pregnancy (*gravidophilia*). Without making any value judgements, it is highly likely that most or all of these addictions have genetic or physiological components which, once properly modified, can greatly reduce or eliminate the addiction if so desired. Many on the list are already suspected to have genetic components, much like *schizophrenia*, drug abuse, *bulimia*, and *alcoholism* (*dipsomania*).

Allergies and intolerances

A food allergy is an allergic reaction to a particular food, although true food allergies are much rarer than is generally believed. In the cases of milk, eggs, shellfish, nuts, wheat, soybeans, and chocolate, sufferers may lack an enzyme necessary for digesting the substance.

In other cases, dust particles, plant pollens, pet danders, drugs, or foods may be allergens for natural IgE-mediated immunosensitivity. Intolerance, a much more common condition, is any undesirable effect of eating a particular food, including gastrointestinal distress, gas, nausea, diarrhea, or other problems. *Urticaria* (hives), *angioedema* and even mild *anaphylaxis* are common reactions to various drugs, insect stings or bites, allergy shots, or certain foods, particularly eggs, shellfish, nuts and fruits. Physical allergies to ordinary stimuli such as cold, sunlight, heat, pollen, pet dander, or minor injury can produce itching, skin blotches, pimples, and hives.

Minor physical annoyances

In a world where most major medical maladies are readily treated, numerous minor medical conditions which today escape our notice will rise up from obscurity and present themselves annoyingly to our conscious minds, demanding attention. These conditions may be of several kinds. First is cosmetics, including small moles, freckles and blemishes on the skin; broken fingernails or unevenly-growing cuticles; minor skin reddenings or pimples; old childhood scars, wrinkled skin, birthmarks or stretch marks; unwanted hair growth in unusual places, or differential hair colour or texture growing in patches; fingerprint patterns that are aesthetically unappealing; and mismatched leg lengths, hands with different left/right ring sizes, an asymmetrical face, or lopsided breasts.

Second is minor aches and pains, which may include headaches; eyebrow hairs trapped in the eyeball conjunctiva; bent-hair pain (*folliculalgia*); dyspepsia; creaking limb joints and stomach growling; ingrown nails and hairs; earwax plugs and temporary tinnitis; chapped lips, canker sores and heat rashes; stuffy nose or gritty eyes upon rising in the morning; dermal chafing marks from elastic bands in clothing; minor flatulence; PMS (premenstrual syndrome); a leg or arm "falling asleep" in certain postures; nervous tics, itches, and twitches; uncracked knuckles, stiff neck, or backache; blocked middle ear following descent from high altitude; restless leg syndrome (*akathisia*); rotationally-induced dizziness, as on an amusement park ride; or rock-and-roll neck, wherein active musical performers or listeners bob their heads violently, rupturing small blood vessels in the neck. Third is minor physical or functional flaws, such as poor stream during male urination, female papillary leakage, colourblindness, snoring, unpleasant body odors, nosebleeds, declining visual or aural acuity, handedness (currently ~90% dextromanual, ~10%

sinistromanual, mild strabismus (eyeball misalignment), bad moods (neurotransmitter imbalances), or post-intoxication hangover.

Undiscovered infectious agents

Peptic ulcers once were thought to result from a stressful life, a purely natural response to a lifestyle choice. Then it was found that the major cause of ulcers is the presence of *Helicobacter pylori* bacteria in the stomach. Bacteria have been implicated in some cases of atherosclerosis and *Alzheimer's disease*, and nanobacteria have been proposed as possible nucleation sites for kidney stones. Other seemingly natural but undesirable conditions may also be due to undiscovered microbial agents, especially since bacteria outnumber tissue cells in our highly infested 20th century bodies by more than 10:1.

Unwanted syndromes

Syndromes are groups of related symptoms and signs of disordered function that define a disease whose cause remains unknown, that is, idiopathic. A good example is irritable bowel syndrome (IBS), which affects up to 20% of the adult U.S. population and includes symptoms of abdominal distention and pain, with more frequent and looser stools. Many are unaware they are afflicted. In 1998 there was no known cause or simple complete treatment for this still "natural" disease. Even more mysterious than IBS is our general activity level—some people seem to have high-energy personalities, while others have more phlegmatic low-energy personalities. Either may be regarded as "natural," but nanomedicine can probably bring this ill-defined neurophysiological variable under human control. The need to sleep is another imperfectly understood syndrome.

It is experienced by everyone and thus was universally regarded as "natural" in the 20th century. Physiological short sleepers were unusual, insomnia or asomnia was thought of as an abnormal state, and there were a few anecdotal but medically undocumented instances of total nonsomnia, such as the celebrated case of Al Herpin.

Psychological traits

Psychological traits which, if identified by a patient as undesirable, might be subject to genetic or physiological modification could include: sexual preference (6-10% of the adult population is homosexual); shyness or boldness; acquisitive or altruistic propensity; misanthropy or philanthropy; theistic or atheistic orientation; loquacity or dourness; childhood imprinting; criminal propensity (up to 1-5% of the population); various recognized personality disorders that affect ~10% of the population such as antisocial, paranoid, schizotypal, histrionic, narcissistic, avoidant, dependent, obsessive-compulsive, and passive-aggressive disorders; panic attacks (experienced at least once by ~33% of all adults each year); and phobic disorders such as social phobias (~13% of the population), specific phobias including fear of large animals (*zoophobia*), snakes (*ophidiophobia*), spiders (*arachnophobia*), needles (*belonephobia*, ~10%), the dark (*noctiphobia* or *scotophobia*), or strangers (*xenophobia*) (total ~5.7%), the fear of blood or *hemophobia* (~5%), *agoraphobia* (~2.8%), and other unusual phobias such as the fears of certain colours (*chromophobia*), daylight (*phengophobia*), girls (*parthenophobia*), men (*androphobia*), stars in the sky (*siderophobia*), the number thirteen (*triskaidekaphobia*), and even the fear of developing a phobia (*phobophobia*).

The above sampling of minor afflictions, almost all considered "natural" in 1998, may come to be regarded as commonplace correctable medical conditions in the nanomedical era. By the time such petty annoyances are deemed worthy of immediate treatment, biotechnology and nanomedicine already will have defeated the most fearsome illnesses of the late 20th century and will have moved on to other challenges. Naturophiles may dissent, but the emerging trend from medical biotechnology is to characterize health, not as a static standard, but rather as a condition defined by the lives that people want to lead. Affirming the volitional normative model of disease, Virginia Postrel concludes:

"Different goals will produce different choices about trade-offs and standards. What makes a condition unhealthy is not that it is unnatural but that it interferes with human purposes. Revering nature [would mean] sacrificing the purposes of individuals to preserve the world as given. It [would require] that we force people to live with biological conditions that trouble them, whether diseases such as cystic fibrosis or schizophrenia, disabilities such as myopia or crooked teeth, or simply less beauty, intelligence, happiness, or grace than could be achieved through artifice.

In a world where it's no big deal to take hormone therapy, Viagra, or Prozac, to have a face lift, or to know a child's sex before birth, a world in which even such radical interventions as sex-change operations and heart transplants have failed to turn society upside down, it is extremely difficult to argue that medical innovations are dangerous simply because they fool Mother Nature."

6

REPLICATORS AND NANOTECHNOLOGY

Diatoms are single celled algae, the 10^5–10^6 species of which create a wide variety of three-dimensional amorphous silica shells. If we could get them to produce useful structures, perhaps by compustat selection experiments (i.e. forced evolution of development or evodevo), their exponential growth in suspension cultures could compete with the lithography techniques of present day nanotechnology, which have limited 3D capabilities. Alternatively, their fine detail could be used for templates for MEMS (micro electro mechanical systems), or their silica deposition systems isolated for guiding silica deposition. A recent paper has demonstrated that silica can be replaced atom for atom without change of shape - a step towards the Star Trek replicator.

Given that diatoms are responsible for ~25% of the world's net primary production they have been the most underfunded organism per unit mass. This might be about to change with a flurry of activity in a new field dubbed 'diatom nanotechnology', which will have its debut in a workshop at the next North American Diatom Society Meeting. Paid and unpaid (so-called '*amateur*') diatomists have spent lifetimes doing research on these single-celled creatures, with perhaps one universal motivation: they are beautiful to behold. Some investigators got carried away, producing art only visible through a microscope, made of arrays of diatom valves.

DIATOM MOTILITY

At first, diatoms were thought to be animals because what are now called *raphid* diatoms move at speeds of up to 25 μm s^{-1} when

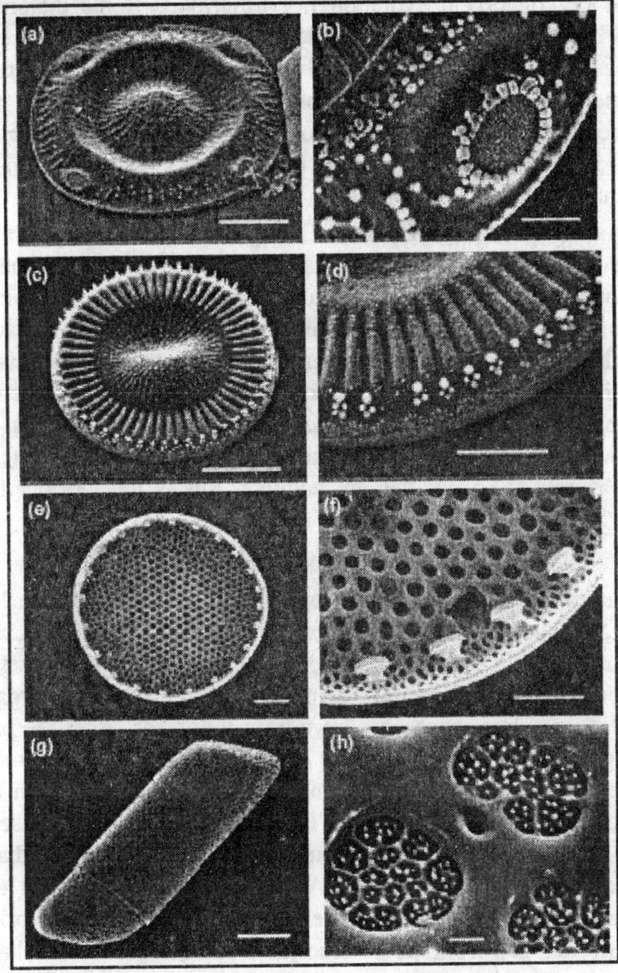

Fig. 6.1. Pairs of scanning electron micrographs showing marine centric diatoms and close-up of surface features. (a,b) Glyphodiscus stellatus; (c,d) Cyclotella meneghiniana; (e,f) Roperia tesselata; (g,h) Isthmia nervosa.

attached to surfaces. They do this with no moving parts, although a 'raphe fluid' is exuded. A slow moving debate has occurred over three decades as to whether the motive force for diatom gliding motility is in the hydration and capillarity of the raphe fluid controlled by actin, or whether actin provides the motive force. A model for myxobacterium motility identical to the capillarity model for diatoms has been proposed, as well as a similar one for cyanobacterial gliding motility (common ancestor or convergent evolution?). Furthermore, some diatoms use 'nozzles' (pore openings) analogous to those on myxobacteria, to secrete an adhesive material, which involves transient motility. Definitive experiments are needed because control of the diatom motor, which can reverse rapidly 'on a dime', somehow detecting a collision or a change in light, is a fascinating problem of potentially universal significance.

Control of motility of diatoms in rigid shells would appear to be simpler than crawling of flexible tissue culture cells of ever-changing shape, and thus a possible key to eukaryotic cell behaviour. Given that diatoms move on surfaces, we could try to use patterned surfaces to guide them into place, perhaps moved along by phototropism or geotropism, where they could become nanotechnology components. In the remarkable colonial diatom *Bacillaria paradoxa* the cells move rhythmically against each other, like a deck of cards sliding back and forth. Because the velocities are additive, this is the world's fastest microorganism.

Diatom Morphogenesis

Diatom shells are composed of amorphous silica but are highly structured. They are basically made of hydrated glass: $SiO_2.nH_2O$. There are 10^5–10^6 species of diatoms, distinguished by their shells. How does an amorphous substance produce such a variety of organized shapes? Two major approaches have been followed: either the pattern forms spontaneously out of silica by *diffusion limited aggregation* (DLA) or there is a prepattern of something else onto which or within which the silica precipitates (which begs the question as to what causes that pattern). Neither approach has seriously invoked the genetics of diatoms, so there is a lot of work ahead. Although diatoms might not tell us how legs and arms and brains of vertebrates are put together, bridging the intellectual gap from the genome (now being sequenced in diatoms) to diatom shell structure would be a great accomplishment. Intermediaries that have been discovered are the silica binding polyamines and silaffins.

Diatom DNA synthesis is silica dependent and a silica membrane transport system exists. Whether these molecular components merely concentrate silica and catalyze its (fractal) DLA precipitation, or somehow form a prepattern onto which silica precipitates, has yet to be determined. They probably account for the 20–40% organic content of diatom silica. The big question, which is also fundamental to debates in evodevo (evolution and development), is whether diatoms (or any organisms) vary '*infinitely*', or are subject to '*developmental constraints*'. Fortunately, we can speed up evolution in the lab, especially for microorganisms, using mutagenic chemicals or ultraviolet light, and assorted methods for selecting those mutants that do best under conditions set. A classical device is the chemostat, which acts similarly to an artificial stomach for microorganisms to grow in: those that grow too slowly are washed out.

It thus selects for speed of growth, amongst those mutants that can use the nutrients that are placed in the incoming medium. A variant of this idea is the compustat, in which a computer decides 'who shall live and who shall die'. The compustat was conceived as a means of pushing diatom morphology to match any preconception we might have. It would work by visually scanning all diatoms in a small growth chamber with a digital camera attached to a motorized microscope, and matching the observed patterns against an ideal 'template'. Those diatoms that were furthest from the ideal would be destroyed with a laser or UV microbeam and the rest subjected to mutagenesis. Hopefully someone will soon build a compustat.

Diatom Nanotechnology

Most fabrication techniques in nanotechnology involve planar lithographic approaches. Three dimensional structures are built up plane by plane, and excruciating acrobatics are required to etch away undercuts and cavities. Diatoms build directly in 3D, which is why they are so attractive for nanotechnology, and of course, with exponential growth in numbers (at one cell division per day), any number of parts we want could be made. The first artificial diatom patterns were synthesized by 3D vapour deposition of silica by Max Schultze back in 1863 (his plates are reproduced), and the feat has only been partially attempted since. If silica is 'merely' precipitated onto an already existing scaffold (prepattern), as some would hypothesize, then we ought to be able to construct artificial scaffolds for silica precipitation.

The physical chemistry of unassisted silica precipitation is well understood. The most remarkable invention in the new field of diatom nanotechnology is that akin to the replicator in the science fiction television and movie series Star Trek. Diatom shells are placed in an atmosphere of magnesium gas at 900°C for 4 hr. Apparently an atom for atom substitution of magnesium for silicon occurs, with no change in 3D shape. Thus silicon oxide, a material of limited usefulness in nanotechnology, is transformed into magnesium oxide. The authors list nine other 'shape preserving gas/solid reactions' that are thermodynamically favoured, and should 'replicate' diatom silica shells into a variety of oxides. The list of uses envisaged includes micro-capsules for medications, water filters (sieves), catalytic substrates, sensors, optical diffraction gratings and actuators, or templates for any of these. As mentioned, diatoms have a peculiar, silica-dependent mechanism for DNA synthesis. This permits silica starvation synchrony to be achieved, by growing the diatoms in Teflon beakers to prevent them from extracting silica from glass.

If pennate diatoms were then killed in mid-formation of their shells, we could produce massive quantities of nanoscale combs, such as are used in comb activators. These could be sorted by size using light scattering in a FACS machine (fluorescence activated cell sorter). Other sources of nanoscale silica objects, although not in such great variety, are the silica bodies found in grasses. One could imagine crops such as wheat being 'genetically engineered' and harvested not only for their seed but also for the silica bodies in their straw. Other, more esoteric organisms that precipitate silica include the radiolarians, sponges and other algae.

The Future

Diatom nanotechnology, conceived as an industrial process, started only as a suggestion in 1988. It is now a highly interdisciplinary, fast moving area, possibly headed for making a major contribution to nanotechnology. Biology as a source of nanotechnology has a few major advantages: large numbers of components are available via the exponential reproduction of the organisms that produce them, and mutation permits us to selectively evolve organisms with the components we want. In the course of pushing organisms to manufacture what we want them to, we are also likely to learn much about whether or not evolution is constrained. Thus bionanotechnology, although driven as an industrial pursuit, might nevertheless make fundamental contributions to our understanding of life.

7

Nanotechnology and Cancer

In 2002, 23% of all deaths in the United States were caused by cancer making it the second biggest killer, only ranking behind heart disease. Every year, over a half million Americans die of cancer and more than a million are diagnosed with the disease. It is also the second biggest killer among children, with nearly 12% of all childhood deaths coming from the disease.

Cancer is a disease in which cells proliferate uncontrollably. Unlike most cells, cancerous cells do not display density dependent growth, meaning they divide with little spatial regulation. Moreover, these cells have the ability to spread by breaking into blood vessels and moving to other systems. Cancer can be fatal due to a combination of its properties. For instance, cancerous cells lose their ability to function normally. That is, they stop responding normally to cellular signals and therefore no longer perform their job. Not only do cancer cells cease working, they also affect neighbouring cells because cell division and metabolism require nutrients and energy; eventually the cells require more nutrients than the body can provide and slowly organ systems begin to fail, a process known as *cachexia*.

Additionally, the growths themselves can cause immense pain or death in hollow organs (such as the *colon*) by blocking the lumen and preventing proper function. Moreover, tumours can cause pressure on the brain which can lead to brain failure, seizures, or partial lack of function depending on the location of the tumour. The formation of cancer requires several genes to be altered through mutations, which can be caused by spontaneous errors in replication or by exposure to carcinogens that alter nucleotides or break the DNA strand. In order for a mutation to lead to cancer, it has to perpetuate the cell cycle. The *cell cycle* is a highly regulated process that ultimately results in the division of one cell into two. In somatic cells, this cycle includes four phases: G1, S, G2, and mitosis (M). During G1 phase, the cell grows as it prepares for DNA synthesis, S phase. Then in G2 phase, the cell grows in preparation for mitosis, in which the replicated DNA is equally divided into two newly formed daughter cells.

Errors in the cell cycle are normally corrected during specific checkpoints at G1 to S, intra-S phase, and S to M. At these points, the cell cycle is temporarily arrested while regulatory enzymes ensure that there are no errors in the DNA sequence. If an error is found, the DNA damage is either repaired or the cell is tagged by a marker protein to commit suicide through apoptosis. If inhibited, the cell cannot properly identify damage, and the cell cycle continues without the appropriate regulation.

The *cell cycle* can be perpetuated through two types of genetic mutations: oncogenes and tumour suppressor genes. Tumour suppressor genes normally are involved with the repair of damaged DNA. Thus, whenever these genes are inactivated, damaged DNA is not properly repaired. According to Ames and Gold (1991), every cell in the body experiences 10^5 DNA damaging events daily. Thus, the regulatory process of repairing DNA is an active and important process. Tumour suppressor genes can be broken down into two categories: *caretakers* and *gatekeepers*. Gatekeepers have a direct roll in controlling cellular proliferation, while caretakers help preserve the integrity of the genome by preventing mutations from occurring. An inactivated caretaker does not lead directly to tumour initiation, but instead it causes genetic instability, which causes subsequent mutations. In contrast, inactivated gatekeepers play a more direct role in the tumourigenesis process.

While tumour suppressor genes are dangerous when inactivated, oncogenes are only hazardous when active, at which point they are capable of inducing cancer in normal cells. Due to this, oncogenes are highly regulated in the body. Additionally, oncogenes have a wide variety of functions. For instance, some encode for growth factors that increase the proliferation of cells, others bind to DNA and regulate transcription, and yet others code for receptors or ligands involved in the cell cycle. If over expressed, however, all of them can contribute to the development of cancer by promoting cell division.

Tumourigenesis, or tumour formation, is a multistep process requiring more than one active oncogene or inactive tumour suppressor gene. If a group of cells has a small number of these mutations, a benign tumour may form. These tumours lack the ability to *metastasize* or spread to other parts of the body. However, if the benign tumour has more mutations, it is possible for it to become malignant.

The process of *carcinogenesis* involves four steps. The first step is *initiation*, in which a carcinogen reacts with DNA causing a strand break or altering a nucleotide to form an adduct. Normally, a DNA polymerase repairs this problem, however, if the DNA replicates before the repair, the error can be permanently fixed into the genome. Most errors of this type have no real effect on the body, but if a tumour suppressor is inactivated or an oncogene activated, the cell has a significant growth advantage, and the next step, *promotion*, may begin. During promotion, a molecule called a *promoter* causes selective proliferation, which may lead to the formation of multiple benign tumours. Through one or more additional genetic alterations, the third step, known as *progression*, may occur. In this step, the tumour cells develop a significant growth advantage, which is so strong that they are able to break through the blood vessel membrane and travel to other areas through the process of metastasis. This actual conversion is the last step, and is referred to as *malignant conversion*. This further establishes the importance of multiple mutated tumour suppressor genes and oncogenes in cancer development. In other words, the growth advantage brought about by one mutation is not significant enough to overcome the natural immunity of the body.

Tumours of this nature are contained because they are unable to break into the blood vessels. However, through multiple mutations, the growth advantage may be increased sufficiently to break through blood vessel membranes. For many years, scientists had no clue how to deal with this growth advantage. As a result, cancer was virtually untreatable, and even today, many types have no specific treatment. Chemotherapy's potential to treat cancer was discovered during December of 1943, when an Allied warship holding mustard gas exploded. As a response to this, the army performed autopsies on the soldiers, which showed that their bone marrow had been destroyed by the gas, thereby inhibiting the production of red blood cells, white blood cells, and platelets. Accordingly, scientists hypothesized that the chemical may be used to fight cancer.

To test this hypothesis, a chemical derived from mustard gas, known as *mustine*, was given to Hodgkin's disease patients and, even in some patients with late-stage Hodgkin's, the disease responded to the drug. In fact, this drug is still a key component of the MOPP (*mustine*, *vincristine*, *procarbazine*, and *predinisone*) regimen, which is one of the two primary treatments for Hodgkin's disease, the other being ABVD (*Adriamycin*, *bleomycin*, *vinblastine*, and *dacarbazine*).

Unfortunately, the treatments commonly used for cancer (*radiation* and *chemotherapy*) are both deleterious to the health of patients, and can actually cause death themselves by weakening the immune system and making patients more susceptible to other diseases. The problem with these treatments is that they are not selective. That is, they act on all rapidly dividing cells causing the most recognizable symptom of cancer treatment: loss of hair. These treatments also inhibit the production of erythrocytes and white blood cells, causing patients to become *anemic* and *neutropenic*. Anemia, a state of insufficient O_2 delivery to tissues, can cause problems with blood clotting, as well as lead to dizziness and lethargy. Neutropenia refers to a decrease in the number of neutrophils in the blood signifying a weakened immune system. When neutropenic, patients are more susceptible to secondary infections; even a common cold can be fatal. Furthermore, chemotherapy triggers neuroreceptors, such as those that bind dopamine and serotonin, which stimulate nausea and cause vomiting.

Not only are chemotherapy and radiation dangerous, they also are not completely effective. According to Dr. Frank Balis, "We attribute our inability to cure many adults with more common forms of solid tumours to the ineffectiveness of chemotherapy to these diseases" (1998). In fact, the average five year survival rate among all cancers in the United States is only 63%. Thus, newer and more effective treatments are being sought by scientists and pharmaceutical companies alike.

In the last few years, the field of nanotechnology has exploded as some scientists believe tiny objects known as *nanoparticles* may be able to help treat a variety of diseases, including cancer. By definition, nanoparticles can range in size from 1 to 100 nanometers. The nanoparticles being studied have a variety of compositions, shapes, and sizes. The most common composition includes either a carbon backbone or the presence of an inorganic metal, such as a gold.

Recently, scientists have discovered that nanoparticles can easily enter cells. However, it is uncertain how this occurs. Dai et al. (2005) claims the influx of nanoparticles occurs by *endocytosis*. In contrast, Bianco et al. (2005) suggest the process happens through insertion and diffusion of particles through the lipid bilayer of the cell membrane. Furthermore and surprisingly, these particles can be linked to proteins, such as antibodies, and still enter cells. Fortunately, cancer cells express certain receptors that are not expressed by normal cells. Thus, nanoparticles attached to antibodies for these receptors can be directed to cancerous cells exclusively.

The ability of nanoparticles to selectively enter cancer cells has duel significance. Firstly, nanoparticles can work as *drug deliverers*. For instance, by linking certain proteins, such as *tumour necrosis factor* (TNF), a protein with known *antitumour* activity, to the particles a new mechanism for fighting cancer can be utilized. Secondly, nanoparticles have been shown to absorb different wavelengths of light than the body, and when exposed to appropriate wavelengths nanoparticles heat up, but the body does not. This method, known as *hyperthermia*, can be used to selectively kill cancer cells by heating nanoparticles that are linked to antibodies.

The specificity of these techniques is key, because unlike the deleterious effects of chemotherapy and radiation, treatment with nanoparticles should result in no major side effects. Furthermore, in preliminary studies, hyperthermia and drug delivery have both been successful, and currently, both hyperthermia and drug delivery are being heavily investigated as treatments for cancer. The purpose of this review is to discuss the nanoparticle techniques of hyperthermia and drug delivery and determine whether they may one day replace the current techniques of chemotherapy and radiation as a treatment for cancer.

Imaging to Detect Cancer Cells

Beyond having the power to treat cancer, nanoparticles may also be used to detect the disease. Moreover, some therapies hope to utilize hyperthermia in such a way that diagnosis and treatment can occur together. There are several techniques scientists are investigating to improve cancer detection and couple it with hyperthermia.

One popular technique involves attaching *bioconjugates*, such as antibodies, to the nanoparticles. Loo et al. (2005) attempted to analyze this technique by utilizing the tendency of breast carcinoma cells to overexpress the HER2 biomarker. Thus, by conjugating an antibody of HER2 to a PEG linker complex, which enhances biocompatibility and blood flow, and then attaching the complex to a gold nanoshell, the particle is linked exclusively to breast cancer cells.

Using this, Loo et al. (2005) cultured three types of cells: cells with the anti-HER2/PEG/nanoshell complex, cells with a non-cancer specific antibody, and cells without nanoshells. These cells were viewed with a dark-field microscope sensitive to scattered light, and only the Anti-HER2 cells showed much light scattering. In contrast, the cells with the non-specific antibody showed some light scattering, but it was not as dense. This illustrates that the Anti-HER2 treated cells attached exclusively to cancer cells, and exposure of light identified cancer cells. Furthermore, when treated with *near-infrared* (NIR) light of around 800 nm, cytotoxicity was observed only in the presence of the cells treated with Anti-HER2 nanoshells. Thus, the hyperthermia treatment was successful, but only with the Anti-HER2 treated cells.

Hyperthermia to Kill Cancer Cells

As mentioned earlier, hyperthermia is the killing of cells through the heating of nanoparticles. One of the problems of hyperthermia is containing the heat in such a way that it does not affect other cells. To combat this, scientists use specific types of nanoparticles for hyperthermia, such as magnetite cationic liposomes (MCLs). These spherical particles contain a positively charged phospholipid exterior that interacts with the negatively charged cell surface, easily entering cells. The inside of the MCLs is a 10 nm magnetite nanoparticle. Additionally, these particles have maintained the ability to bind to antibodies and can provide tumourspecific contrast enhancement.

Gene Therapy/Hyperthermia Combination

Hyperthermia appears to be effective in some cases by itself, however, in advanced stages of several types of cancer, such as melanoma, it may not be sufficient. Furthermore, to treat cancer, hyperthermia requires many treatments. However, in conjunction with other processes, scientists hope to find a way to use one round of hyperthermia to eradicate the disease. The combination therapies revolved around the use of substances to boost anti-tumour immunity. Thus, in addition to hyperthermia, the cancer cells will be assaulted by a revamped immune system. Ito et al. (2003a) have been analyzing the use of one such protein, heat shock protein 70 (HSP70), in conjunction with hyperthermia with MCLs. Expression of this protein protects cells from heat-induced apoptosis, but recently, it has also been shown to be a key component in immune reactions.

To analyze HSP70 gene therapy combined with hyperthermia, Ito et al (2003a) analyzed how mice with malignant melanoma reacted to tumours that had been given a plasmid containing human-inducible hsp70 complimentary DNA. The primary finding was that hsp70 gene transfer successfully boosted the immune system during hyperthermia. They determined this by comparing tumour size after exposure to hsp70 containing plasmid, hyperthermia, and the combined treatment. Both treatments alone showed improvement, but in each case, additional treatments would be required because the tumours began to grow again at around the tenth day. The combined therapy, however, completely eradicated cancer in 3 of the 10 mice with only one treatment. Because hyperthermia can be used multiple times without any negative effects, it is believed that the cancer could have been eradicated in the other mice with subsequent treatments. Moreover, tumours with the combined therapy were 16 times smaller than the hyperthermia only treated tumours after thirty days, and 24 times smaller than the tumours given hsp70.

Hyperthermia with Dendritic Cell Addition

The use of immune triggering proteins is not the only way to boost anti-tumour activity. For instance, mature dendritic cells (DC) are an integral part of a normal immune response, which stimulate the growth of $CD4^+$ T cells, $CD8^+$ cytotoxic T lymphocytes, and natural killer cells. Unfortunately, mature DCs cannot take up antigen, and thus addition of these cells would not result in the proper immune response. Injection of immature DCs, however, has been reported to cause antitumour activity.

Tanaka et al. (2005) decided to go straight to the source by actually adding additional dendritic cells (DC) after mouse EL4 T-lymphoma tumours were treated with hyperthermia. While only 1 in 8 of the mice treated with hyperthermia alone had complete tumour regression, 6 in 8 of the mice treated with hyperthermia and immature DCs had complete tumour regression. Based on this, it appears the tumour cells killed by hyperthermia release antigen proteins which the immature DCs take up and are then presented to T cells via MHC class I and/or II antigens.

Drug Delivery Using Nanoparticles

Drug delivery is the carrying of drugs using nanoparticles specifically to the cells causing the disorder. In the case of cancer, these drugs are frequently known *chemotherapeutic* agents. Intravenously, these drugs cause a variety of side effects. However, by linking them to nanoparticles the drugs go directly to the source and do not affect healthy cells. As is the case with hyperthermia, certain types of nanoparticles are better adapted for drug delivery than others. For instance, nanoparticles composed of colloidal gold easily attach various drugs. Colloidal gold is a dispersed solution of nanoparticles of Au^0. Additionally, polybutyl cyanoacrylate (PCB) nanoparticles attach drugs, protect them against enzymatic degradation, reduce their toxic effects, and limit distribution of the drug outside the target area.

Tumour Necrosis Factor and Colloidal Gold

Tumour necrosis factor (TNF) is a cytokine that affects coagulation, lipid metabolism, insulin resistance, and proper function of endothelial cells. It is produced during immune response primarily by monocytes and macrophages and has the ability to induce death in tumour cells. Unfortunately, TNF causes systemic toxicities that have prevented it from being used as an anti-cancer drug. This toxicity can be attributed to rapid uptake of TNF by the *reticuloendothelial system* (RES). Through the use of colloidal gold nanoparticles, Paciotti et al. (2004) were able to construct a vector which can avoid detection and clearance by the RES. Thus, the nanoparticles (PT-cAu) delivered TNF specifically to tumour cells, eliminating the associated systemic toxicity.

Next, Paciotti et al. (2004) compared treatment using native TNF and PT-cAu-TNF which showed both reduced tumour size in a concentration dependent manner. However, mice given 12 μg native TNF suffered 25% fatality and all given 24 μg native TNF died whereas none of the mice treated with PTcAu-TNF perished. Furthermore, Figure 2b illustrates that while 15 μg of Native TNF has approximately the same affect on tumour size as PT-cAu-TNF through 16 days, the survival rate using the native form is 40% lower. Thus, without the colloidal gold nanoparticles, TNF is extremely toxic. These nanoparticles help TNF circumvent the RES and enter selectively into cancer cells, which ultimately causes tumour cells to die.

Localized Chemotherapy

As mentioned earlier, the main problem with chemotherapy is that it is not tumour specific. Thus, chemotherapy drugs tend to act on all rapidly dividing cells. Through the use of nanoparticles, however, the same drugs can be linked specifically to cancer cells at higher concentrations for longer periods of time. Thus, the drugs not only have increased cytotoxic activity, but also adverse side effects are limited.

Doxorubicin

Doxorubicin hydrochloride (Dox), also known as *adriamycin*, is a cytotoxic anthracycline that is an essential component of chemotherapeutic regimens used to treat acute lymphoblastic leukemia, breast carcinoma, Hodgkin's and Non-Hodgkin's lymphoma. The drug works by halting DNA replication, and thereby preventing further proliferation of the disease.

Fortunately, Dox's anti-tumour activity has been widely documented, and there is no reason to think it would behave differently if attached to a nanoparticle. At the same time, intravenous treatment of Dox causes systemic toxicity that can cause severe diarrhea, neutropenia, anemia, hair loss, and heart damage. Thus, scientists are investigating the use of different types of nanoparticles that can be used to deliver Dox directly to cancer cells, ultimately preventing systemic toxicity.

Reddy and Murthy (2004a) investigated this by analyzing two different polymerization techniques for making polybutyl cyanoacrylate (PRC) nanoparticles: *dispersion polymerization* (DP) and *emulsion polymerization* (EP). The result of each polymerization technique produced structurally similar molecules. The difference, however, was that the EP nanoparticles were smaller. Therefore, Reddy and Murthy (2004a) sought to find out whether the size difference of the PRCs affected the nanoparticles' ability to deliver Dox. They found that EP particles provided a longer half-life of Dox in the blood and a lower tissue distribution, which is consistent with their previous finding that EP nanoparticles have enhanced permeability and retention effects. Conversely, DP nanoparticles were quickly cleared into the RES. Both techniques demonstrated a significant increase in bioavailability of Dox compared to intravenous injection of Dox solution. Together, the experiment identified the EP nanoparticles as a potential method of improving Dox therapy by reducing systemic toxicity.

Following the polymerization study, Reddy et al. (2004b) examined the affect of Doxorubican loaded poly(butyl cyanoacrylate) (DPBC) nanoparticles on Dalton's lymphoma. They found that the DPBC nanoparticles sequestered in the tumour after subcutaneous injection much better than did free Dox. Additionally, they noted that there was a low amount of Dox found in the heart from the DPBC nanoparticles, and confirmed that Dox delivered by DPBC nanoparticles has an increased retention time within tumours. This confirms the results of the previous experiment, and also shows that cardiac toxicity may be limited through this technique.

Ma et al. (2004) developed another type of nanoparticle to be used for Dox delivery to tumour cells. The particles, known as *carbon magnetic nanoparticles* (CMNP), were created using a new technology known as *dense medium plasma* (DMP) technology. The particles consist of a carbon-based host structure with iron and iron oxide particles evenly dispersed. The CMNP-Dox and intravenous free Dox were applied to osteosarcoma cells to test anti-proliferative activity. The results showed that at the highest dose, free Dox had no significant effect on the tumour cells compared to CMNP-Dox, which completely stopped proliferation at 120 μg/ml Dox. Interestingly, at 240 μg/ml, CMNP-Dox had a reduced effect, believed to be because of steric hindrance caused by excess nanoparticles. One of the chief advantages of this system, however, is that it can be made in one step under atmospheric pressure using inexpensive chemicals, such as benzene and acetonitrile, making it both effective and cost efficient.

Paclitaxel

Paclitaxel is a chemotherapy drug that can be used to treat Kaposi's sarcoma and metastatic breast, ovarian, and bladder cancer. It is an anti-microtubule compound that prevents continuation of the cell cycle and thus proliferation. In the case of bladder cancer, doxorubicin and mitomycin C are ineffective treatment options due to their inability to pass through the transitional epithelium in the wall of the bladder known as the urothelium. Since paclitaxel is lipophilic, however, it can freely pass through the urothelium. The FDA approved formulation for paclitaxel includes the solvent *Cremophor*.

Cremaphor causes paclitaxel to become entrapped in the micelles of the bladder, which lowers the drugs ability to penetrate the urothelium. To combat this, Wientjes (2003) used DMSO as a surface-active agent that disrupted Cremaphore micelles and enabled paclitaxel to be delivered to the tumours; however, this technique caused increased urine production and associated drug removal. Consequently, with less time in contact with the cancerous cells, paclitaxel was less effective.

Wientjes's et al. (2004) second attempt to facilitate the transfer of paclitaxel through the urothelium utilized gelatin nanoparticles loaded with the drug. These nanoparticles are hydrophilic and thus uptake fluid rapidly allowing for paclitaxel to be released easily. This is important because the quicker the drug is released, the longer its exposure to cancer cells before urination.

The concentration of paclitaxel in the urine, which was collected during treatment, was 2.6x that of the cremophor/EtOH formula. Additionally, 87% of the drug was released in two hours, compared to only 45% after 3 days for paclitaxel-loaded poly(ethylene oxide)- poly (lactide/glycolide) nanospheres used to regulate smooth muscle cell regulation. In summary, paclitaxel loaded gelatin nanoparticles were able to penetrate the urothelium of the bladder and rapidly release the drug, making them a promising treatment for bladder cancer.

Gene Delivery using Nanoparticles

Nanoparticles can deliver proteins with anti-tumour activity into tumour cells and additionally, they can be used to deliver chemotherapeutic drugs directly to tumours, avoiding systematic toxicity. The versatility of these small particles also allows them to transport plasmid DNA with tumour suppressor genes to tumour cells. This causes a tumour suppressing protein to be produced which induces tumour cell apoptosis, effectively fighting the cancer.

MDA-7

First identified in human melanoma cells, the human melanoma differentiation associated gene 7 (*mda-7* or *IL-24*) is a tumour suppressor gene. In late stage human melanoma, MDA-7 protein is absent, whereas in early stage melanoma it is present. Accordingly, this gene product is likely involved with progression of the disease. Furthermore, the protein is absent in a variety of human tumours including lung, breast, and colorectal carcinomas and sarcomas, and thus, it is believed to be involved in both the development and progression of these human cancers.

Previous studies have shown that through using adenoviral vectors, expression of MDA-7/IL-24 triggers cytotoxic related cell death and growth suppression in several human cancer cells. Moreover, normal cells are not affected by exposure to *mda-7*gene, making it a potentially strong anti-tumour therapy. In 2003, Chada et al. used an adenoviral receptor to deliver *mda-7* to tumours in the lungs. The results were promising, because this procedure caused expression of MDA-7 induced apoptosis in the tumours. Unfortunately, the adenovirus vector can cause an immune response and liver toxicity. Therefore, a new vector for *mda-7* delivery to disseminated cancers is needed.

Ito et al. (2003c) demonstrated that DOTAP: cholesterol nanoparticles can transport tumour suppressor genes to tumours in the lungs and increase the transgrene expression of these genes. Based on this, Ramesh et al. (2004) tested the use of cationic DOTAP: cholesterol (Chol) nanoparticles as a vector for delivery of *mda-7* gene. They found that cells treated with the DOTAP/mda-7 gene showed significantly fewer tumours. Additionally, they found no resistance to multiple treatments with this therapy, as well as no systematic toxicity. Furthermore, the treatment was still successful in immunodeficient and immunocompetent organisms. Thus, using DOTAP: Chol nanoparticles as a vector for the *mda-7* gene is a novel approach for cancer therapy that shows much promise (Ramesh et al., 2004).

Discussion

In this chapter, three promising techniques for treatment of cancer using nanoparticles have been chronicled: hyperthermia, drug delivery, and gene therapy. These techniques each have several advantages over the current treatments of radiation and chemotherapy. Firstly, neither of these treatments causes systematic toxicity. In fact, both hyperthermia and drug delivery can be directed specifically to cancer cells. Ultimately, this is advantageous because it greatly reduces the physically and psychologically demanding side effects of chemotherapy and radiation, which include, but are not limited to anemia, neutropenia, hair loss, diarrhea, sterility, and nausea.

These side effects are thought to be worthwhile because of chemotherapy's effect on cancer, but all cancer cells are not responsive to chemotherapy. Furthermore, some cancers develop resistance to chemotherapeutic drugs. There are several reasons for this. As mentioned in the beginning, tumour cells have a variety of mutations, and all tumour cells do not have the same mutation. Some mutations allow cells to randomly develop resistance to drugs because they no longer express the protein receptors to which the drug interacts. Thus, the cells without the receptor have a growth advantage, and if another drug is not used, these cells will proliferate rapidly. Additionally, tumour cells may produce more target proteins than the drugs can bind. Since chemotherapeutic agents are not specific, the concentration of the drugs cannot be raised, as other systems of the body would be effected as well. Furthermore, enhanced amplification of the *MDR1* (Multiple Drug Resistance) gene results in the encoding of a large transmembrane protein which can stop certain drugs from entering a cell and also eject drugs already in it.

With chemotherapy, any form of resistance requires another type of drug; however, nanoparticles may hold the key to circumventing such resistance. Early trials with hyperthermia and gene delivery show that each technique may be used multiple times. Hyperthemia, for instance, does not work on hindering processes inside the cell, but instead, it heats the cell up to such high temperatures that it denatures proteins and DNA. Heat shock proteins that stabilize proteins to prevent denaturing are themselves denatured when exposed to heat of this magnitude. Thus, hyperthermia can be done repeatedly without detrimental effects to other systems or the threat of tumour cells becoming resistant to it. Moreover, through gene delivery, expression of tumour suppressor genes inside tumours can be controlled, so in essence, tumours are forced to fight themselves. Early experiments suggest, that using the mda-7 gene in this manner can be performed repeatedly and cells develop no resistance.

While hypothermia and gene therapy appear to circumvent resistance, through the use of higher drug concentrations and exposure time, localized drug delivery provides another option. The use of nanoparticles allows for higher concentrations of drugs, such as native tumour necrosis factor and doxorubicin, to be used. This is possible because the nanoparticles specifically target cancer cells, and thus, there will be no associated systemic toxicity. Because a higher amount of the drugs can be used, the initial treatment has a larger effect, as more of the drug is able to interact with the tumour cells. Additionally, a second treatment can be administered much more rapidly afterward, since the rest of the body does not have to recover. Together, there is a much smaller chance that the tumour would develop resistance, because treatment can take a shorter period of time. However, it is possible that some of the tumour cells have innate resistance, in which case, another drug would have to be used.

In summary, the techniques of drug delivery and hyperthermia using nanoparticles have the potential to decrease side effects while increasing the cure rate of cancer patients. These techniques promise a substantial improvement over chemotherapy and radiation. Over the next few years, if the research conducted on nanoparticles continues to find promising results, the treatment of cancer all over the world may be substantially altered. The cure for cancer may in fact be close at hand.

Treatment of Cancer with RNA Nanotechnology

Using strands of genetic material, *Purdue* University scientists have constructed tiny delivery vehicles that can carry anticancer therapeutic agents directly to infected cells, offering a potential wealth of new treatments for chronic diseases.

The vehicles look nothing like delivery trucks, though that is their function once inside the body. Instead, these so-called *nanoparticles*, which are assembled from three short pieces of ribonucleic acid, resemble miniature triangles. The microscopic particles possess both the right size to gain entry into cells and also the right structure to carry other therapeutic strands of RNA inside with them, where they are able to halt viral growth or cancer's progress. The team has already tested the nanoparticles successfully against cancer growth in mice and lab-grown human cells.

"RNA has immense promise as a therapeutic agent against cancer, but until now we have not had an efficient system to bring multiple therapeutic agents directly into specific cancer cells where they can perform different tasks," said research team leader Peixuan Guo, who is a professor of molecular virology at Purdue with joint appointments in Purdue's Cancer Research Center, School of Veterinary Medicine and Weldon School of Biomedical Engineering. "Physicians have hoped that nanotechnology might provide a solution to the problem, and it's possible that the application of these tiny triangles could lead to the solution."

"With these devices, Dr. Guo was able to deliver three different therapeutic agents into a cell at the same time," said Jean Chin, a scientist at the National Institute of General Medical Sciences, which is part of the National Institutes of Health. "This is an incredible accomplishment that points to the versatility and potential medical value of these nanoparticles."

The research appears in two related papers being published in the scientific journals Nano Letters and Human Gene Therapy. Guo's team created their nanoparticles by linking together different kinds of RNA, a task that their previous research has given them ample opportunities to practice. Several years after building a tiny "motor" from several strands of RNA that mimic those in a bacteria-killing virus called phi29, the team learned how to manipulate these stringy molecules into different shapes, including rods, triangles and arrays.

"We speculated at that time that these shapes would be useful purely as physical scaffolding on which more sophisticated nanodevices could be constructed," Guo said. "But RNA, which carries genetic messages within cells, also has many therapeutic functions. We realized that if we built different kinds of therapeutic RNA onto the RNA scaffolding and created a single structure, we might be able to respond to several challenges that have confronted the medical field."

RNA molecules come in many variant forms, and the sort that the team mimicked from the phi29 virus - called *pRNA* - also can be linked to other types of RNA to form longer, hybrid strands with properties the researchers could assign.

"We looked around for RNA strands that would behave in certain ways when they encounter a cancer cell because each of them needs to perform one step of the therapy," Guo explained. "An effective agent against cancer needs to accomplish several tasks. It needs first to recognize the cancer cell and gain access to its interior, and then it needs to destroy it. But we'd also like the agent to leave a trail for us, to mark the path the molecule has taken somehow. That way, we can pinpoint the location of the cancer and trace the outcome after the treatment."

To accomplish these tasks, the team turned to other forms of RNA that can interfere with the goings-on inside cells. The team sorted through a variety of RNA forms that have shown promise for disease treatment and found three that could perform each of the desired tasks. One example is "small interfering RNA," or *siRNA*, which deactivates certain genes in cells. The others are RNA aptamers, which bind to cancer cell surface markers, and ribozymes, which can be designed to degrade specific RNA in cancer cells or viruses.

"We linked each of the three therapeutic strands with a piece of pRNA, forming three hybrid strands," Guo said. "Then, using techniques we learned from our earlier work, we were able to combine all three into triangles that are between 25 and 40 nanometers wide. This is the Goldilocks size for any nanoparticle that is to be used in the body - not too big, not too small."

Particles larger than about 100 nanometers are generally too large to pass through cell membranes into the cell's interior, Guo said, and the body has a hard time retaining particles smaller than 10 nanometers. But the tiny triangles fit, and they worked well enough to interrupt the growth of human breast cancer cells and leukemia model lymphocytes in laboratory experiments.

"One characteristic of cancer cells is that they do not stop growing, which is one reason tumours develop," Guo said. "Once inside, the siRNA essentially instructs the cells to 'stop not stopping.' The nanoparticles had done their work on the breast cancer cell cultures within a few days."

Additionally, the team found that the nanoparticles completely block cancer development in living mice. A group of mice that were in the process of developing cancer were tested with the nanoparticles, and they did not develop the disease. A second group that was tested with mutated inactive RNA all developed tumours.

"The results are very promising, but we still have several hurdles to jump before we can test this therapy on people," Guo said. "First and foremost, we must ensure that it is as safe as we think it is. Some RNA can be toxic to noncancerous cells as well, and though our nanoparticles appear to go straight to the cancer cells where we want them to go, we have to be sure they do not go anywhere else before we can inject them into a living person." Stability of the RNA also is a factor the team must consider. Although they previously published data indicating that phi29 RNA nanoparticles are more stable than other RNA, Guo said the team still needs to find better ways to protect the RNA from degradation by enzymes in the body. Although the group still needs to prove the safety of their tiny creations, Guo said, they remain confident that their work is a milestone for medical nanotechnology.

"Many studies have shown that therapeutic forms of RNA, such as siRNA or ribozymes, could be put together to kill cancer, but the main obstacle has been finding the delivery method that can bring them to specific cells simultaneously," Guo said. "Nanotechnology is beginning to pay off here in that it may have provided us with a solution to the problem. We hope to enhance the work we have done so far and refine it for human trials."

8

NANO-NETWORK

Although there is a broadening social interest in the development of a powerful and general nanotechnology, the public discourse to date has largely avoided a comprehensive examination of its social dimensions, focussing instead on what is and is not scientifically possible. In this regard, much attention has been paid to the feasibility of Richard Feynman's famous 1959 vision, i.e., whether it is possible to manufacture complex molecules *atom-by-atom*. Whether Feynman's hunch is correct – that "it would be, in principle, possible for a physicist to synthesize any chemical substance that a chemist writes down" – has been fiercely debated in the scientific literature and the popular press.

The most famous version of this debate. The recent point/counterpoint exchange between Richard Smalley and Eric Drexler, illustrates a deep division within the nanotechnology community. The above compilation of *soundbytes* from the public exchanges between Smalley and Drexler over the past few years is not meant to provide full coverage, nor even a summary of their scientific positions. Quite to the contrary, these remarks were very purposefully selected to demonstrate that there are other things at play besides the testing of hypotheses – that this is not *mere* scientific discourse.

While this debate has been extremely influential within scientific circles, it is suggested that such discussion is not particularly useful in the broader policy arena. In our view, despite their good intentions, this is not the best way for prominent scientists to assist in the development of appropriate regulatory structures for nanotechnology. This type of rhetorical exchange is not the best enabler of sound policy and planning. Although the development of sound social policy about a given technology must certainly commence with considerations about what is presently foreseeable, in this brief article we suggest that it is also important to contemplate possibilities that are not necessarily congruent with today's forecasts. We further suggest that scientific forecasting is itself an insufficient social safeguard against a technology said to have the potential to revolutionize our ability to control and manipulate matter. As an alternative, we propose, policy makers ought to embrace a foresight model that aims to develop a broader network of social participants in their deliberations about the future regulation of nanotechnology.

Mend the Gap

Policy-making is inherently a challenging task – a task made more difficult when faced with future uncertainties, In the face of rapid change, it is not good enough to simply debate about what we think is and is not scientifically possible today. Nor is it sufficient to state that "[t]here is no scientific evidence to support the notion that nanoparticles and nanotubes – the main components of many nanotech-based products – pose risks on human health and the environment. While such statements, if true, are an important claim in advancing the argument that the perceived risks of nanotechnology are likely to be overestimated and overrated by mass media and the like, the policy debate does not and ought not to end with the conclusions of our science *de jour*. Rather, we must learn how to co-ordinate science and technology policy so that we can plan for alternative futures. This will involve broadening the debate beyond physicists, chemists and engineers.

As the authors of a recent report noted: As the science of NT leaps ahead, the ethics lags behind. Activist groups have appropriately identified this gap, and begun to exploit it. We believe that there is danger of derailing NT if serious study of NT's ethical, environmental, economic, legal, and social implications does not reach the speed of progress in the science. Minding the gap is indeed an important first step. Mending it, however, is the more challenging next step. In this section we briefly describe a well-known alternative to the point/counterpoint discourse approach that has been adopted by Smalley and Drexler. By reiterating this alternative approach, we hope to remind those interested in the ethics and science of nanotechnology that there are other discussions to be had.

Rather than focussing primarily on competing scientific visions about the feasibility of molecular manufacturing, we hope to connect that discourse to existing techniques that have been used in other fields to identify and assess the bridge between our possible futures and the present. What we are promoting is what one future studies author has described as: *a code to communicate between social actors in science, technology and society* ... a combined analysis and communications process in which informed parties and stakeholders participate in a forward-looking exercise to identify the most important issues in the emerging S&T portfolio.

Foresighting, as it is sometimes called, is a methodology for examining the long-term future and finding answers for the present as a means of guiding technology policy. It represents an historical shift from short-term to long-term thinking; from past-oriented to future-oriented; from linear to non-linear 'system' thinking; and from an either/or to a multiple option mindset. It has been denned as: ...the process involved in systematically attempting to look into the longer-term future of science, technology, the economy and society with the aim of identifying the areas of strategic research and the emerging of generic technologies likely to yield the greatest economic and social benefits. This approach involves "[a] process by which one comes to a fuller understanding of the forces shaping the long term future ... which should be taken into account in policy formulation, planning and decision, making."

'Foresighting' can be distinguished from 'forecasting'. Forecasting is the passive attempt to diagnose or predict future events. Smalley's seminal *Scientific American* article, for example, merely forecasts that self-replicating nanobots cannot and will not be part of our future. Conversely, foresighting aims to actively change or create the future by linking it to the present. It focuses on the challenges of tomorrow, today. Thus, "the major difference between foresight and forecasting is that in forecasting the conclusions for today are missing." The process of foresighting is premised on the assumption that the future is not fixed and that alternative futures exist."

Foresighting can be used in various ways. According to Slaughter, there are four major applications: "(i) assessing possible consequences of actions... (ii) anticipating problems before they occur ... (iii) considering the present implications of possible future events [and] (iv) envisioning desired aspects of future societies." As the literature points out, foresighting as a tool for "decision-shaping' rather than "decision-making' offers many benefits including: engaging policy-makers and experts in actively planning for the future, identifying potential problems early, verifying expectations and examining trends, bringing people together to create a suitable future, strengthening existing networks, and educating the public on urgent future-related issues. Foresighting could have a positive impact on nanotechnology policy by providing a means for analyzing its broader social and economic implications.

While some believe that nanotechnology has the potential to eliminate the problem of resource scarcity, others have pointed out that a technology which allows that 'anything can be made from anything' is sure to have an impact on our ecological systems. Similar considerations will arise in the context of economics. For instance, unless nanotechnology offers a solution to the problem of inflation, we should not necessarily assume that near costless materials' production will necessarily result in decreased prices.

A foresighting methodology is needed to commence an assessment of nanotechnology's potential impact on these and other core socioeconomic structures. For example, certain visions of nanotechnology, idealized, could lead to significant economic disruption. Substantially revised or perhaps even alternative economic systems might one day be required to ensure that the fruits of nanotechnology (like some of the information technologies that preceded it) are not short-lived. Similar considerations might be necessary to avoid a proliferation of existing disparities in wealth and power, and the creation of new divides between the haves and have-nots. All of these things indicate that we need to further develop a set of methodologies that will help us to identify and assess the bridge between our possible futures and the present.

Building a Broader Nano-network

Scientific forecasting, conceptual modelling and the testing of hypotheses in the laboratory – though they are all key to a bright future – cannot provide sufficient social safeguards for a science said to have the potential to revolutionize our ability to control and manipulate matter. Mending the gap requires the development of a broader nano-network.

Instead of standing on the sidelines, cheering on a combative and adversarial scientific arm-wrestling match, diverse groups of social actors ought to assemble to examine potential profits and pitfalls of the technologies that miniaturize from as many different angles and perspectives as possible – with the aim of consensus building. As one Australian professor put it, true foresighting requires us to build an '*epistemic community*' founded on "a number of principles around which the community members inter-subjectively construct a consensus." These principles would include agreed-to methods and models for assessing and understanding causal relationships, common language and jargon, and political values concerning the policy implications and what policy choices should be preferred. Although some foresighting techniques currently employed in other contexts rely primarily on experts, many believe that a more complete methodology ought to include a broader range of participants from the social sciences, the humanities and the arts.

The general public can and ought also to play a role in understanding and analyzing the social implications of various foresighting activities. This type of active and inclusive participation not only generates excellent opportunities for public education and consultation about possible future events, but also enhances an expectation that "the rationality as well as the legitimacy of political decisions can be improved." As Cuhls and Grupp point out, Discursive approaches make for more rationalized discussions, because they focus on the need to provide arguments. They introduce reasons as a standard for political discussion.

Therefore, they correct the strategic (party) intellectuality and argumentative propaganda which is common in the public (mass media) confrontations. In contrast to the domain of experts - where it is possible, advertently or inadvertently, for researchers to promote their own ideologies, interests and agendas through the language of science - extending the nano-network to include laypersons and experts from relevant non-scientific disciplines would allow for greater political transparency. It might also promote a more informed and actively engaged public whereby "collective knowledge and the efficient performance of all actors in society and their capability to exchange information result in a steering resource similar to power or money. An approach that creates a broader nano-network, involving other social actors in discussions and decision-making about the future regulation of nanotechnology, would enhance legitimacy and foster public trust.

Conclusion

In the quest for knowledge, scientists, unlike elected officials, are not held responsible for safeguarding the public interest. They are not generally obliged to explore issues that extend beyond their own research interests, nor are they required to consult with members of the public or others working beyond their own domains of expertise. Although most scientific policy is the product of negotiation and contestation, such policies should not be determined primarily on the basis of contestations amongst scientists directly involved in the scientific or technological break-through in question. Just as progress in science is sometimes overshadowed by politics, policy choices surrounding the adoption or regulation of a particular science or technology can become clouded by the rhetorical assertions of particular scientific stakeholders.

In the absence of a 'social contract' between scientists, government officials, and the public, heated exchanges between scientists must not become a policy maker's preoccupation. Whatever place rhetoric might have in science, on its own, it is an ill-suited method for policy analysis - especially in fields where there is little consensus but great uncertainty. This is an increasingly significant consideration when one recognizes that scientific discourse is often used as a means of building powerful though divisive social networks. In the face of competition it is usually those scientific networks that are the most successful in translating their own interests on the largest scale that have the greatest impact on how a new technology is developed and implemented.

Likewise, the most powerful scientific networks can also have an impact on how such technologies are eventually regulated. With this in mind, we ought to be very careful not so foster a divisive nano-network. It is suggested here that, in the face of scientific uncertainty, we ought to be oriented towards building a broader, more inclusive network that embraces actors from diverse sectors and enables the development of an overlapping consensus in the shaping of future policy.

9

Nanotechnology and Dialysis

More than 800,000 patients suffering from *end-stage renal disease* (ESRD) are today treated with renal replacement therapy (RRT), including renal transplantation. The dialytic forms of RRT include hemodialysis and peritoneal dialysis, with the former being used in a large majority of patients in most countries.

Extracorporeal hemodialysis has undergone a series of developments in the last three decades mostly related to improvements in biomaterials and enhancement of the design of the critical disposable devices and machines. In particular, while the initial hemodialysis systems were developed piecemeal by dedicated artisans, who were also engineers and physicians, the most recent devices and machines are highly sophisticated and rely on large production series and standardized manufacturing procedures. This evolution has taken place in conjunction with a significant reduction of hemodialyzer dimensions. An important part of this process has been the significantly better understanding of the physiological effects and consequences of dialysis at a micro-scale level, leading to adequate blood circuit design, smooth and biocompatible blood pathways, standardized thickness and porosity of membranes, and accurate alarm systems and failure controls.

As a result of this enhanced understanding and technical sophistication, we operate today with hemodialyzers with minimal blood priming required, constant and reproducible performance, and minimal dialytic losses of protein compounds, even in the presence of significantly high sieving coefficients for large molecules. One may argue that as we continue to increase dialyzer efficiency, we are getting close to the limits imposed by available or potential membranes and by the commonly utilized mass separation processes. For these reasons, a new interest is growing in hyper-selective barrier separation processes made possible by specifically designed synthetic membranes. Similarly, selective solid separation processes based on the use of newly designed adsorbent materials are under evaluation for newer and more efficient blood purification techniques.

The evolution of biomaterials and dialytic techniques has paralleled the evolution of other technologies, including computers and biotechnology. Computers have evolved beyond any expectation, increasing speed, memory, and analytical capabilities, while decreasing in physical size. From the large main frames whose operational time had to be shared among scientists of different institutions, hardware has evolved to highly efficient personal computers, hand-held devices, and potent microchips with enormous capacity for data storage and management. Computers have also become key tools in accelerating processes that are occurring in the field of biotechnology, while different biological processes are providing new models for the development of new hardware and software (i.e. DNA-based computers). This process of interaction between biology and electronics/mechanics has been defined 'bionic convergence' and may result in a tremendous impact on medicine. In February 1997, researchers created Dolly, a lamb cloned from the DNA of an adult sheep. What was considered impossible in the past, proved feasible, and demonstrated the growing power of new biotechnologies. Further examples could be represented by new DNA-based vaccines, new generations of drugs or newly conceived bioartificial organs. This evolutionary process benefits from a series of recent discoveries and scientific advancements deriving from the merging of new technologies and materials. In this context one of the exciting and emerging fields, with applications in engineering, physics, and biomedicine, is nanotechnology.

NANOTECHNOLOGY

Nanotechnology has recently become a buzzword in the popular press: it has been used indiscriminately to describe everything from micro-machines and tiny chemical blobs to the latest shrinking circuits on computer chips. However, according to Eric Drexler, the word 'nanotechnology' should only be used to describe atomically precise functional machine systems developed on the scale of the nanometer (one billionth of a meter: 1 nm = 1/1,000,000,000 m). A vanguard of medical explorers is exploiting the tools of nanotechnology to manipulate molecules either in biological or mechanical systems that may contribute to cell life and death or illness and health. This approach plays on the size scale of biology itself. Theoretically it should be possible to design tiny tools to safely and effectively repair tissues or biomaterials, just as a mechanic works on a car's engine using tools that are on the same scale as the engine. This might sound like science fiction, but it is reaching the verge of possibility because of the above-mentioned 'bionic convergence'. The human genome project, diagnostic 'labs on a chip', hundreds of molecules sorted and characterized by special procedures, represent just a few examples of this evolving field.

To simplify, nanotechnology is the science of manipulating and building materials on the nanometer scale. Nanomedicine may be defined as the monitoring, repair, construction and control of human biological systems at the molecular level using engineered nano-devices and nano-structures. Assuming these actions will become feasible, the number of possible applications for medical purposes will be countless.

Table 9.1. Potential application and pathways for nanomedicine

Early detection of illness accomplished through regular, inexpensive biomonitoring
Accurate diagnosis of genetically defined subtypes of diseases
Biochemically tailored nutraceuticals, antibiotics, immunotherapy
Artificial blood products, skin cartilage and other tissues including possibility for hybrid bioartificial organs
Gene therapy with cellular repair and protein manipulation
Cell-repair machines, healing machines, tissue reconstructors and artificial immuno systems are definitely subject to the assembler-based nanotechnology

The full promise of nanomedicine, however, is unlikely to arrive until after the development of precisely controlled or programmable medical nanomachines and nanorobots. Scaling down mechanical devices was first proposed by Galileo. The possibility of further scaling down to microscopic machines was suggested by the Nobel-winning physicist Richard Feynman in 1959 and later was described at length by Eric Drexler in his popular books 'Engines of Creation' (1986), 'Unbounding the Future' (1991) and finally in his more recent technical book 'Nanosystems: Molecular Machinery, Manufacturing, and Computation' (1992). When such nanomachines are available, the ultimate dream of every healer, medicine man and physician throughout recorded history, to prevent and/or cure disease at the organ/cellular/molecular level, will, at last, became a reality. Programmed nanomachines will be able to execute curative and reconstructive procedures in the human body at the cellular and molecular levels. The dream of scientists already working in this field is even to generate the 'ultimate nanomachine' or the micro-machine with capability of self-replication and generation of new machines: this is what the experts call a 'universal assembler' and what some believe could someday bring science closer to the very origin of life.

Potential of Nanomedicine in Renal Diseases

What is the practical potential for applying nanotechnology to the field of renal disease? How might nanotechnology be used to prevent renal damage, to monitor renal dysfunction, and finally to replace renal function in ESRD patients? Nanotechnology might contribute to early diagnosis of renal dysfunction with a series of implantable micro-sensors that could detect early biochemical abnormalities or incipient flogistic processes. Once ESRD has been reached, these sensors might help to 'drive' conservative or substitutive therapies such as hemodialysis. If the possibility to manipulate and repair biological systems and tissues at molecular levels is successfully exploited by nanotechnology, not only repair of the glomerular basement membrane might be possible, but even early removal of immune complexes or dense deposits from the subendothelial or sub-epithelial spaces of the glomerular loops may be possible to prevent subsequent pathological lesions from forming. Finally, selective immunotolerance for implanted renal grafts might be created by using circulating nanomachines designed to modulate the immunological response in transplanted patients.

Does Nanotechnology Apply to Dialysis?

Since nanomachines cannot yet be built, it is necessary to credibly establish that such devices are feasible and that molecular manufacturing and manipulation are consistent with known laws of physics and obey sound engineering principles. There are indeed several precursor technologies such as telemicrosurgery, tissue engineering and DNA clonation. The first step in this field was made when a team of scientists at the Almodom Laboratories of IBM used a tunneling scanning microscope to move xenon atoms into patterns on a crystal surface. They were able to arrange individual atoms to build an image of the IBM corporate logo. This was the first public demonstration of precisely manipulated matter at the atomic level.

At this stage of development, if nanotechnology is defined more broadly to include not only assembler-based machine systems, but also any construction of molecular structures sufficiently complex to function as machines or devices, then prototypes of some initial applications already exist. Biosensors based on precisely designed synthetic membranes with ion channels of 1.5 nanometer of diameter or micro-devices for personal status monitoring are in fact under evaluation by the Defense Advanced Research Projects Agency. But how far or close are these applications to dialysis? In the year 2000, a retreat took place with experts in nanotechnology meeting with a group of scientists and physicians involved in dialysis to brainstorm about the potential applications of nanotechnology to dialysis. An interesting discussion originated from the analysis of the unmet clinical needs of today's RRT.

New possibilities were identified for application of nanoscience to current devices; precise manipulation of biomaterials and surfaces to make them perfectly hemocompatible; miniaturization of various components and devices to create portable or wearable artificial kidney systems; micro-sensors for continuous monitoring and biofeedback. But while future theoretical applications are being discussed, some initial practical examples of nanotechnology are appearing on the stage. Fresenius Medical Care has recently launched its latest membrane named Helixone. This is a polysulfone-based high flux dialysis membrane in which the porosity of the inner layer is finely controlled at the level of nano scale. Polysulfone membranes may have different solute and water permeability depending on their thickness, porosity and structure. In particular, the structure is characterized by an internal thin skin layer (1 μm) surrounded by a sponge-like layer generally asymmetric with a higher density of the polymer in the more internal region. The effective permeability of this membrane is given by the internal skin layer and its porosity.

The porosity of a membrane (expressed by the number of pores per unit of surface area and the curve of pore size distribution) must be optimized in dialysis in order to achieve the maximal permeability for solutes between 0 and 40,000 D with a minimal or no leak of albumin (MW 48,000 D). As described earlier, three hypothetical membranes characterized by different porosities and consequent sieving characteristics. Membrane A presents an average pore size smaller than the others and therefore the sieving capacity is limited for solutes of medium-high molecular weight range. Small solutes are still effectively cleared since the number of pores is sufficiently high. This represents a typical low-flux hemodialysis membrane such as cellulose-based Cuprophan. Membrane B, on the other hand, presents a greater

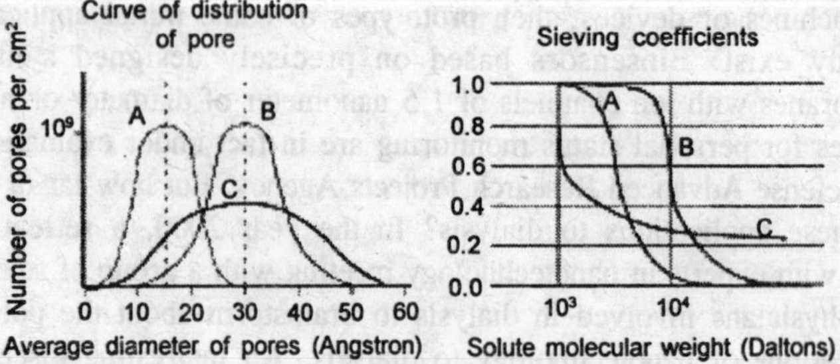

Fig. 9.1. Schematic representation of three different membranes in terms of number of pores, pore size distribution and solute sieveing coefficient.

average pore size, although it maintains an adequate number of pores. This describes a typical well-designed high-flux synthetic membrane usable either in hemofiltration, high flux dialysis or hemodiafiltration. Finally, membrane C presents a typical albumin leaking membrane with a wide distribution of pore sizes but a limited number of pores. This is a good approximation of what is occurring across the human peritoneal mesothelium.

The ideal progression in terms of pore size, diameter and configuration from the old generation of high-flux membranes to the more recently improved polysulfone membranes and finally towards the optimal pore configuration in an ideal high-flux membrane. In detail, while in the old generation of membranes the skin layer presented a series of pores very different in size and structure leading to fluctuations in performance, the more recent high-flux polysulfone

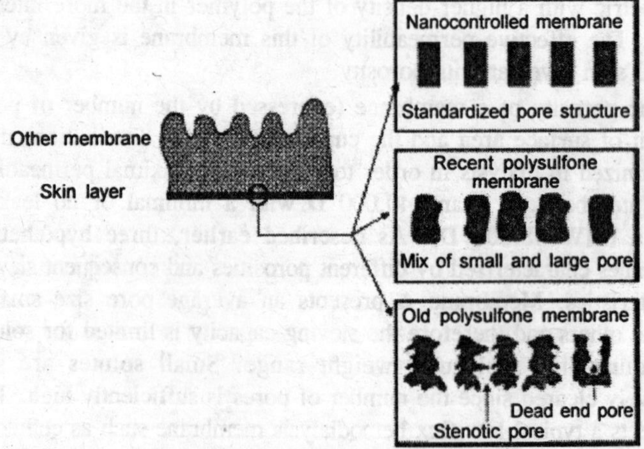

Fig. 9.2. The structure of the skin layer characterizes the performance of the membrane.

membranes display more homogeneous pore size diameters and a higher number of large pores. In such conditions however the transport of small solutes is still limited by the number of pores per square centimeter and increased average diameter of the pores is accompanied by a certain leak of albumin due to the presence of some pores exceeding the desired cut-off of the membrane. In the ideal membrane on the contrary, the performances are maximized by the increased number of pores and the transport of larger molecules is enhanced thanks to the average increase in pore diameter. At the same time however, the fine control of the pore size and structure allows for a precise range of pore diameters reducing the leak of albumin to negligible values. These results are clearly reported and characteristics of the new Helixone ® (FMC) membrane are compared to the already optimal characteristics and performance of the PS600® (FMC) polysulfone membrane. The nanocontrol spinning (NCS) procedure applied to the production of the Helixone membrane produces a significant effect on the structure of the skin layer at the nanoscale level. The number of pores is increased, while the spectrum of pore diameters is narrowed and concentrated around the desired values. While the sieving value for medium sized molecules such as β2 microglobulin has been increased in comparison to the previous membrane, the sieving for albumin has been reduced with virtually no leak of albumin at all. This finely tuned control of membrane porosity allows one to obtain an almost ideal condition for a dialysis membrane. In practice the nanocontrol spinning (NCS) procedure applied to the Helixone membrane has made possible a refined control of the number,

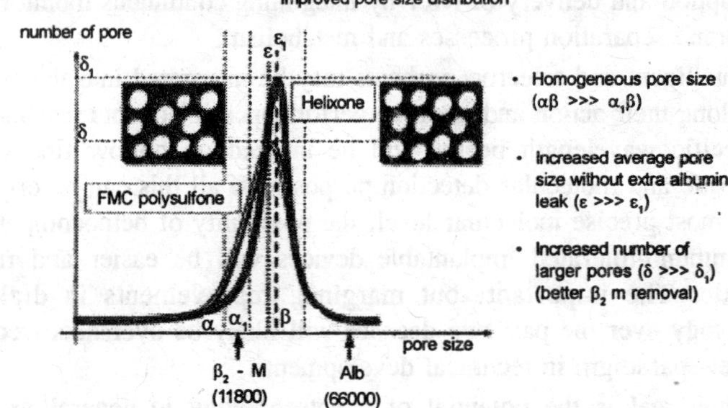

Fig. 9.3. Schematic representation of the improvements obtained on the Helixone membrane in term of number of pores (δ to δ_1), pore size distribution ($\alpha\beta$ to $\alpha_1\beta$) and average pore size (ε to ε_1).

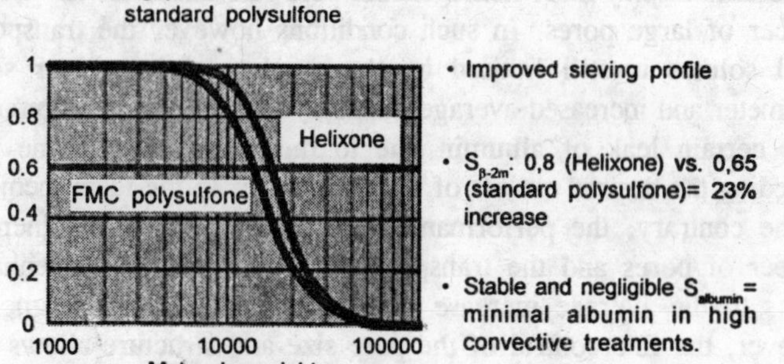

Fig. 9.4. Effects of nanocontrol spinning on the sieving coefficients of Helixone compared to the previous polysulfone membrane.

diameter and anatomical characteristics of the pores. By this molecular manufacturing procedure, nanotechnology is applied to dialysis membrane manufacturing leading to improved performances and safety. This approach represents an initial step in the application of nanotechnology to hemodialysis. It also highlights the enormous potential of molecular control in biomaterials design and development at a nanoscale level.

CONCLUSIONS

This paper has provided a brief overview of nanotechnology and its present and future applications to the field of RRT. What do we see as the future of nanotechnology and RRT? We will definitely see a progressive miniaturization of current dialysis machines, monitors and devices. We will create increasingly 'intelligent' fibers with immobilized enzymes and diagnostic tools able to provide diagnosis, prescription and delivery of RRT by integrating continuous monitoring, membrane separation processes and metabolism.

Intelligent and selective sorbents may be integrated into the fibers to prolong their action and improve performance. Nanodots responding to specific wavelength beams will be utilized on hollow fibers for diagnostic and molecular detection purposes. If all this can be created at the most precise molecular level, the possibility of hemocompatible and immuno-tolerated implantable devices will be easier and more realistic. The important, but marginal improvements in dialytic technology over the past two decades will likely be overshadowed by this new paradigm in technical development.

How real is the potential of nanotechnology in general and in biomedicine in particular? Just surf the web for a few minutes and see the huge number of companies and institutions investing in this field, and developing training programs and career paths in nanotechnology. Academic as well as business positions are being created at a fantastic speed. Just remember: Leonardo da Vinci clearly envisaged the flying machine much earlier than the Wright brothers took off for the first time. Knowledge and research will help make possible in the near future what today is simply considered impossible. The dream of today will be the reality of tomorrow.

10

IMPLICATION OF NANOTECHNOLOGY

What might be the prospects for the commons in a future of expanding cyberspaces, material abundance, and long life? Thinking about the implications of nanotechnology research and development offers a stimulating way to explore some of the challenges and opportunities facing management of common property resources in the new millennium. *Nanotechnology* is a strong candidate to drive a future wave of change in technology and society, following and amplifying waves of change from information technology and biotechnology.

THREE ROUTES TO NANOTECHNOLOGY

Development of molecular machinery is scientifically feasible, though decades of scientific and engineering work may be needed before it is widely applied. At least three routes lead to nanotechnology, defined as precise control over the construction of molecular systems at the nanoscale of billionths of a meter. The computer industry is investing heavily in learning how to build smaller and smaller devices, shrinking manufacturing towards the atomic scale, (the *top-down route*). Biotechnology is discovering how to tailor the molecular machinery of DNA and proteins, (the *wet route*). Manipulation of individual atoms with atomic force microscopes foreshadows molecular assemblers which could build materials atom by atom, (the *bottom-up route*). Nanotechnology will probably be developed, one way or another. The challenge for speculation, and for technical and social invention, is how it may be used.

Potentials

The potential impacts of nanotechnology are immense. Shrinking computer components to atomic scale could enable computers to continue to grow cheaper, smaller, and more powerful for decades hence. Tiny nanomachines could monitor and make repairs inside cells, curing disease and extending life.

Molecular assemblers might build materials to order, making matter as controllable and easily reproduced as software, while also disassembling wastes and pollution to recover elements and compounds for reuse. How these potentials are developed, and who has access to them, depends on social institutions. These institutions include not just private ownership and state regulation, but also collective action in the management of common property, old and new commons.

Nanotechnology Scenario

An accelerating rate of technological change may lead to a singularity, a point beyond which it may be impossible to even foresee what might follow. However this paper suggests that efforts to create and manage common property will persist, however drastic the changes which may occur. Scenarios are useful tools for formulating ideas and policies that will be more resilient in the face of inevitable surprises. Nanotechnology, and particularly the vision of nanotechnology laid out in Eric Drexler's book, *Engines of Creation*, offers a feasible scenario of possible changes in technological capacity. Considering the implications of nanotechnology challenges some of the current conceptions and preoccupations in the study of common property. The point of this paper is not to make specific predictions, but to explore some possible consequences of technological change, using the example of nanotechnology, and to identify some possible implications for the study and management of common property.

Organization of the Paper

The first section of this paper looks at "open source" approaches to the collective production and management of intellectual property in a commons of freely shared information resources. Molecular engineering may transform the use and reuse of natural resources, and the second section raises questions about the implications of abundance for management of common-pool resource stocks and flows. Nanotechnology may further expand capabilities to create shared spaces, both physical structures and cyberspaces.

The third section notes how stakeholders reconstitute the governance of shared spaces as quasi-commons, mixing modes of private and public property with the influence of stakeholder voice and exit. The fourth section outlines how health, wealth and long life might bring increased attention to long-term endeavors, encouraging the crafting of common property institutions for millennial projects.

Open Source in Information Commons

Most studies of common property have focused on common pool resources, such as forests, fisheries, rivers and pastures in which a group of users share rights to use the resource. For such common pool resources, one person's use subtracts from the amount available for others to use, so users are rivals to some degree. By contrast, information can usually be duplicated at little or no cost, without necessarily diminishing its value to existing users. This potential for non-rival use and sharing is a distinctive characteristic of intellectual property. Conventionally, intellectual property rights, embodied in patents, copyrights and legally protected trade secrets, have been seen as necessary incentives to promote the production of knowledge, especially commercially useful knowledge. Such views emphasize the need to reward individual creativity, and tend to neglect the cumulative and collective processes through which knowledge is created. The emergence of open source software illustrates how an alternative approach to sharing information resources has enabled and accelerated the creation of a new commons.

Open Source Software

The Linux computer operating system has emerged as a serious challenger to the near-monopoly of Microsoft Windows. Linux exemplifies the strengths of collective action by a swarm of volunteer programmers to create freely available software. While Linux may be the best known example in the wider world, it is only one of many such efforts, ranging from the EMACS word processor pioneered on mainframe computers to the Apache software running the majority of web servers. Open source approaches have had surprising success in overcoming problems of coordination, diseconomies of scale and scope, which had been seen as inevitable constraints on large software projects, even well-funded programming projects managed by large corporations such as IBM or Microsoft. Eric Raymond's essay, "The Cathedral and the Bazaar" analyzed the differences between the centralized planning of conventional software, prone to delays and errors (*bugs*), with the decentralized vigor of open source approaches for creating sophisticated, reliable programs.

Open source software has characteristics of common property, in being freely available for use, nowadays posted on the internet where anyone can download it. It is the collective product of a community of programmers working together, most of them helping refine only a small piece of the overall project. Certain individuals may have "*ownership*" rights, based on initiative and community reputation, to decide what changes will be included in new releases of the software. Raymond argues that reputation is the key incentive underlying the post-scarcity "*gift culture*" of computer programmers participating in open source efforts, competing for status by giving things away.

As with other forms of common property, open source communities face conflicts concerning how to institutionalize access to common property. Intense debates have raged about how to provide a legal basis for open source software which will preserve accessibility while maintaining incentives for continued development. Legally the software is not in the public domain, licenses give rights that allow others to use and modify the software. Licensing provides a legal mechanism for defining the boundaries of the group of legitimate users. In practice, membership is largely constituted through the electronic mailing lists used to discuss and announce improvements.

Programming has strong roots in academia, partaking of scientific traditions for sharing knowledge. However open source software has proven able to thrive in commercial domains where one might usually expect proprietary software to be the rule. Raymond argues that a key reason for the economic viability of open source software comes because the vast majority of software is actually not sold directly, but instead written for internal use in producing goods and services. The businesses that pay the programmers who write the software are not concerned to maximize sales value. Instead, those businesses, and others who use open source software, are more concerned about reliability, quality, compatibility with popular standards, availability from multiple sources, and other factors, among which cost is often a relatively minor consideration. In some cases, companies may use a strategy of freely releasing software which enables them to profit from sales of related equipment or services. Open source strategies build on the potentials for easy sharing of electronic information, emerging through individual and organizational incentives that encourage communities of software writers and users to create and share new information commons.

Open Source for Nanotechnology

The increasing ability to precisely control the assembly of matter at the molecular and atomic level (one definition of nanotechnology) promises to make matter like software. Building equipment, food and other materials might become as easy, and cheap, as printing on paper is now. Just as a labourious process of handwriting texts was transformed first into an industrial technology for mass production and then individualized in computer printers, so also the manufacturing of equipment and other goods might also reach the same level of customized production. If "*assemblers*" could fabricate materials to order, then what would matter would not be the materials, but the design, the knowledge lying behind manufacture.

The most important part of nanotechnology would be the software, the description of how to assemble something. This design information would then be quintessentially an information resource, software. The same principles which have promoted the rapid development of accessible, affordable and innovative open source software might also be applied to the creation of knowledge for nanotechnology. Nanotechnology could maintain the paradigms of openness, public criticism and development as a community effort which are already part of its foundations in current nanoscience, rather than assuming that proprietary secrecy is the only route to further development. This could accelerate research and development, and promote accessibility, while also promoting safety.

Opening up Intellectual Property

Raymond argues that the advantages of closed source come through profiting from specific secret information, and that what has not been adequately appreciated is the extent to which the advantages of open source may far outweigh the potential profits of such proprietary information. He suggests that five factors may favour the choice of open source:

We can expect that open source has a high payoff where

(a) reliability/stability/scalability are critical, and
(b) correctness of design and implementation is not readily verified by means other than independent peer review. ...
(c) when the software is a business-critical capital good [due to the desire not to be dependent on a single supplier] ...
(d) in software that establishes or enables a common computing and communications infrastructure. ...
(e) when key methods (or functional equivalents) are part of common engineering knowledge.

These criteria can be applied to nanotechnology. For nanotechnology, reliability and stability are crucial. For computers scalability means being able to expand to deal with more users and more information, and somewhat similar concerns arise if designs are used to produce multitudes of tiny nanomachines. Peer review may be essential to detect and correct errors. Companies using nanotechnology to produce other goods and services may prefer open source, to avoid dependence on a single supplier who keeps software secret. Common standards and infrastructure would promote cumulative development of the technology.

It is not yet clear the extent to which nanotechnology may rely on commonly known procedures or unique techniques, but most of the foundations may well be developed as openly shared scientific knowledge. It would seem that the rationale for open source would apply strongly in the case of nanotechnology, particularly if the principal products will not be the production equipment, assemblers, but rather designs for producing materials and machines, i.e. software. An obvious starting point would be to promote open source approaches in developing modeling software for nanotechnology design. Such design is already being done on an exploratory basis, largely in a scientific mode, to see what sorts of molecules might be build for nanomachines.

Creating commitment to open source in modeling software for the longer term, even after commercialization, would establish a foundation for accessible information. This would gain the advantages of peer review to examine design safety issues. It would build a coalition of individuals and organizations interested in supporting open source software for modeling nanotechnology designs. Such an initiative would benefit from first-mover advantages, and could establish standards, creating the intellectual infrastructure for a new field.

A further rationale for open source, especially when viewed from a wider societal perspective, comes from the much lower transactions costs of managing an open information commons, compared to the higher costs of establishing and protecting private intellectual property. Open source offers an alternative approach, applicable in at least some interesting cases, where it may well be in the public interest to promote (or at least avoid discouraging) common property institutions for information commons, rather than relying only on private forms of intellectual property.

Strategy of Openness

Public sharing of information could also help to deal with some of the hazards that may accompany nanotechnology. Initially, and by analogy with biotechnology, concerns about the safety of nanotechnology focused on containment, avoiding nanomachines that might run wild and dissolve the world into "*gray goo.*" However over time it has become clearer that safety can be built into the design of nanomachines so they are easily shut down, and dependent on specialized inputs (*artificial vitamins*). Deliberate design could make nanomachines far more controllable than the viruses and bacteria used in biotechnology. The risk of nanotech going wild could be made as unlikely as worrying that automobiles will go feral, stealing gasoline and reproducing themselves in the wild. The dangers that would remain would be ones of deliberate abuse, scenarios of hackers, terrorists and warfare. These risks are real and need to be addressed. The methods already developed to cope with atomic, biological and chemical technologies could help to deal with these risks, but open source approaches might make some further contributions.

One of the arguments for open source software has been that a big and diverse group of people can quickly identify flaws, whether bugs or weaknesses that might be misused. Open source advocates argue that collective efforts to develop robust software, resistant to hacker vandalism, outpace individual or proprietary efforts. In the case of nanotechnology, safety protocols could be built into the core of designs. If these designs have the first mover advantages of widespread backing, then those who might wish to abuse the technology would be at a great disadvantage, with little choice but to use software which already has safety designed in from the beginning.

In additional to internal design characteristics such as making the technology failsafe, failsoft and dependent on scarce and controlled inputs, a key requirement might be transparency to external monitoring in key applications, particularly for any technology capable of self-replication. Openess to inspection is already a key part of international regimes for controlling atomic, biological and chemical technologies. An emphasis on transparency would use a strategy of openness. Rather than emphasizing secrecy and restriction of information to the police and other government authorities, the strategy would be one of transparency, making monitoring information available to whoever is interested. Enforcement then need not be limited to narrow efforts by police, but would draw on the interests of a broad community, including media as well as commercial and political competitors with a strong interest in detecting and deterring illegitimate use.

Patents represent a well-established form of intellectual property which compromises between the goal of providing incentives to innovators, while requiring them to make their new knowledge publicly available and limiting the period where the inventor controls rights to use of the new ideas. Similarly, environmental concerns have brought increasingly strict requirements that industries make publicly available information about what substances they release into the environment. It seems reasonable to expect that regulation of nanotechnology might reach similar compromises between private and public interests, including strong requirements that key applications of nanotechnology be subject to open monitoring.

Open Source and Biotechnology

It may be interesting to consider the extent to which open source principles could apply to the case of biotechnology. Genetic engineering obviously has evoked strong concerns about safety, for which reliability and stability of the technology are crucial. Like nanotechnology, biotechnology is very concerned with the ability to produce its products billions of times over. Peer review may play a similar role in enhancing biotechnology, and safety concerns might reinforce preferences for open approaches, rather than proprietary secrecy. Companies using biotechnology to deliver other services, such as health care, rather than just selling genetic products (i.e. seeds) might well prefer open source suppliers. It is less clear how the principle of a common infrastructure might apply, though demonstrating compliance with safety standards might be important. Many forms of biotechnology rely on well-known processes, found in biology rather than being invented, while others make more use of "unique or highly differentiated services. It would appear that open source approaches might be relevant for many aspects of biotechnology, although there would likely still be major areas where proprietary approaches would retain advantages.

At a minimum, open source approaches might offer some useful options concerning the regulation of intellectual property rights related to genetic resources. The example of computer code shows how collective open source efforts can thrive, even while knowledge is produced and used within commercial activities. Similar principles might apply concerning genetic codes. Rather than a purely proprietary or purely public approach to intellectual property in genetic resources, a more differentiated approach might be feasible. Such an approach might facilitate collective action to discover and make available genetic knowledge in ways that would be difficult through purely public or private approaches. In at least some cases, proprietary rights may be neither necessary nor even the best way to promote innovation, while open source approaches could facilitate broader access to useful knowledge about genetic resources.

Information resources differ from common pool resources since information can be easily shared and duplicated, without subtracting from its availability to the original user. Both common- pool and non-rival resources may sometimes benefit from institutional arrangements that allow a diverse group of users to share in the production and governance of the resource, in ways which differ from public or private ownership. The principles of openness and cooperative effort which have promoted the growth of open source software might also foster the growth of nanotechnology design information, and assist in promoting safety, transparency (monitorability) and access to biotechnology and nanotechnology.

IMPLICATION OF ABUNDANCE

Much thinking about management of the common property resources such as water, forests and fisheries is driven by assumptions of increasing scarcity. Assumptions of increasing scarcity may underlie arguments that resources must be privatized in order to internalize externalities, create incentives for better management, and prevent a "*tragedy of the commons.*" Students of common property have challenged such arguments, emphasizing the difference between commons where access and aspects of usage are regulated, versus "*open access*" resources with no such controls on how the resource is used. Research has documented the capacity of resource users to jointly organize common property institutions that regulate access, mediate conflicting claims and sustainably manage shared resources. The potential economic value of the resource is a major factor, inducing conflict, raising incentives for private appropriation, and possibly justifying the transactions costs involved in defining private rights over a resource such as water, fisheries, forest or range land. In contrast to prevalent scenarios of increasing scarcity, the potentials of nanotechnology might instead lead to increasing abundance, and increased capability for environmental regeneration. This shift in relative resource costs could then influence choices between the three broad institutional alternatives of public, private and common property resource management, in ways that might well favour common property institutions.

Elementary Recycling

The ability to precisely control the assembly of matter at the molecular level also implies the capacity to disassemble materials, purify air and water, clean up pollution and possibly transform waste dumps into sources of valuable materials. Most recycling, of glass, paper or plastic for example, currently goes on at the chemical level of complex molecules and bulk materials. The presence of undesired elements and compounds often makes recycling difficult or uneconomic. Nanotechnology could make it easier to extract and reuse desirable atoms and compounds, and sort out temporarily unneeded elements for storage or use elsewhere. Properly deployed, nanotechnology could be a green technology, reducing the need for environmentally damaging resource extraction, enabling sustainable resource availability, and contributing to environmental restoration and regeneration.

From Flow to Stock

Water purification offers an example which can be used to explore possible implications of using nanotechnology for purification, and consequent changes in resource costs. Carbon nanotubes might not only enhance water purification technologies to the point where they far outcompete reverse osmosis or distillation in arid areas, but even become preferable options for water treatment in humid regions, at least for any place not too distant from the sea. Water is normally managed as a flow resource, used once and then passed on, whether quickly along a stream or more slowly through recharge and extraction in aquifers. Rivers and seas tend to be treated as sinks, into which water can be dumped after use.

Tertiary wastewater treatment plants already can yield water with higher quality than the original sources used for urban water supplies. Many of the innovations in reducing industrial water use involve closing cycles, reusing water many times, or even indefinitely, with only occasional replenishment to compensate for minor losses such as evaporation. Fully implementing closed systems for urban water supply would likely face opposition due to attitudes about cleanliness and purity, especially for drinking water. A change might be more feasible with systems operating on "*graywater*" principles, distinct from water used for cooking, washing and bathing. Applying closed-cycle approaches to industrial and domestic water supply would mean that water in such systems would be managed primarily as a resource stock, rather than a transient flow. If implemented widely, such changes might dramatically reduce dependence of cities on upstream water sources. On the one hand this might reduce pressures to acquire new water sources, while on the other hand it might reduce the interest of downstream users in supporting improved watershed management upstream. Technological change would not only expand the envelope of technical possibilities, but could lead to changes in key characteristics shaping how the resource is managed, in this case, a shift from management of one-way resource flows to management of recycled resource stocks.

Atmospheric Commons

Improvements in purification technology might also drive changes from a "use it once and pass it on" approach for *indoor air*. *Outdoor air* quality in wealthier countries has already improved dramatically, much of this driven by regulation of air as a public resource. Indoor air quality is a more recent concern, and struggles over its management are shown, for example, by changing social norms and legal restrictions on smoking cigarettes. Concern for energy conservation has already driven much greater attention to improving insulation and reducing unwanted air loss in temperate climates, and much potential for further such improvements exists, reducing flows relative to stocks. Nanotechnology would represent an ultimate purification technology, capable of removing radon, formaldehyde and other such pollutants, and making total recycling at least conceptually feasible, as well as the elimination of unwanted emissions. Increasing ability to recycle air within a building might lead air to be managed even more as a stock, with the flow element of the resource playing a smaller role. Private business, and common property organizations such as condominiums, would gain greater capacity to manage the resource themselves, reducing (for better or worse) their dependence on public management of air quality. At the same time, eliminating emissions and recapturing pollutants such as sulfur and carbon dioxide could become much more feasible and affordable. Depending on costs, this might reinforce or undermine schemes for tradable emission rights. How such changes might play themselves out is hard to predict, but emphasizes the importance of not assuming that choices between state regulation, common property management and market approaches will always face the same structure of costs and technologies.

Paper Problems

The paperless office has so far proved elusive, with paper consumption increasing, though recycling has also risen. Improvements in the ability to create and restore useful molecules could further reduce the need to harvest wood or other natural fibers. Other technological innovations, such as display screens which look and act like paper, would have similar effects. Reduction in demand for wood products would reduce one incentive for state control or private ownership of forests. Where control over forests is contested, this might allow more space for retaining communal forms of ownership, able to regulate diverse users without splitting up the resource into individualized units. However, demand for "natural" materials may well persist or increase, even if commodity prices continue to decline. Resource abundance is unlikely to eliminate struggles over control of forests and other natural resources, but could reshape the economic value of the resources at stake. Struggles over forests, seashores, lakes, mountain valleys and other areas might be increasingly driven not by resource extraction, but by other forces, particularly the demand for positional goods such as resorts and second homes.

Diamonds and Water

Economists are fond of using the comparison between diamonds and water to illustrate how economic value depends on scarcity. Water is essential to life, but relatively abundant, while diamonds are not essential, but scarce and prized. The ability to recycle materials at the elemental level could well transform scarcity into abundance. Similarly, the choice between private property, common property and state institutions for managing resources is influenced by the value of resources and the transactions costs of different institutions for managing them. Over a range of resources, the potentials of nanotechnology could enable recovery and reuse of materials which are currently disposed into the environment, reducing demands for extracting resources from natural sources. Elementary recycling may drastically reduce demand on natural resources, and so reduce their economic value. It may reduce the incentives for individuals and states to seek control over such resources. In some cases, common property institutions might offer lower transactions costs, equitable access, and more effective means for a community of users to jointly manage a shared resource. The declining value of private benefits, compared to the stakes a broader community of users have in the resource (*externalities*), thus might induce a relative shift from away from private proprietorship. In many cases, this could come through increased assertion of public state rights over private owners, mediated by government action. However, in some cases this might, also lead to efforts by specific groups of users to create new commons, following contemporary examples such as land trusts and wetlands protection.

Inclusive Entitlements

Cheap does not mean free. Abundance does not assure access. One of the cruelest problems of the contemporary world is the persistence of famine, in an era of abundant food production, set amidst broader juxtapositions of poverty and prosperity. Amartya Sen (1981) analyzed how famine comes not from lack of food, but from lacking the capability to obtain food. Food is present in times of famine, but too many people lack the entitlements, the socially established command over resources, that would enable them to obtain food. Similarly, the technological potentials created by nanotechnology will not automatically be available to all. The previous section argued that open source approaches could contribute to the accessibility of nanotechnology.

The institutional arrangements through which people become capable of obtaining goods and services would still play a powerful role. Common property institutions offer further lessons about how access can be shared, while still managing resources sustainably. Such principles might be used to organize access to the fruits of material prosperity, and those resources which are still scarce. Between the exclusiveness of private ownership, and the bureaucratic limitations and flawed incentives of state provision, common property institutions may offer one way to expand access to abundance. As with earlier commons, membership in institutions organized on such principles might offer inclusive entitlements, bringing both capabilities and responsibilities.

Innovation and Abundance

Even without nanotechnology, there are strong grounds to hope for a future of increasing relative abundance for natural resources, rather than the conventional presumptions of increasing scarcity. Julian Simon's book, *The Ultimate Resource* offers both theoretical explanations for innovations that promote resource efficiency and substitution, and abundant empirical documentation of declining commodity prices, increasing abundance relative to the cost of labour and capital, and of the ability to sustain such trends. From another perspective, in *Natural Capital*, Paul Hawken, Amory Lovins and Hunter Lovins outline creative, "*green*" innovations available for multiplying efficiency and drastically reducing resource use. From either perspective, the potential exists for sustaining the world's growing population at levels of material comfort that exceed those of currently affluent nations, while protecting and restoring natural habitats.

The prospects for nanotechnology further reinforce the arguments for continued technological innovation which, if wisely applied, can generate resource abundance, and facilitate environmental restoration. It then becomes even more important to consider the role common property institutions might play in the management of resources such as water, forests and seas, especially where the economic gains which might be extracted by individual proprietors are low relative to the externalities affecting other stakeholders who have strong concerns about management of commons or other shared spaces.

Constituting Quasi-commons

Nanotechnology may bring increasing capacity to cheaply construct and reconstruct the built environment of physical spaces, as well as expanding the diversity and accessibility of cyberspaces. Much thinking about management of traditional commons has been rooted in conceptions of egalitarian communities, small face-to-face groups cooperating democratically to managed a shared local resource. However, in thinking about new commons and other shared spaces, it may be important to reconsider the relevance of familiar assumptions about commons, communities, how roles in governance are constituted, the relative importance of voice versus exit, and the interaction of pluralistic communities. A few examples may help introduce the complexities of quasi-commons:

1. The resort town of Seaside, Florida is one of the landmarks of the "New Urbanism," a school of architects designing attractive mixtures of spaces for living, working, shopping and playing. The park at the center of Seaside is marked by a semicircle of shops, framing a green commons, where children play and people can gather to listen to concerts, watch movies, and enjoy other community events.
2. In suburban shopping malls, walkers often gather in the morning to exercise in a safe, climate controlled space. Political campaigners and others have sought to ensure that they can speak, distribute leaflets and exercise other rights of speech on the nominally private grounds of malls. Teenagers often make malls a place of their own, (if not inhabiting the electronic spaces of chatrooms or networked videogames).
3. In early 2000, there was an uproar that a prominent agency placing advertising on websites might be gathering names, e-mail addresses and other personal information about website visitors without their knowledge or permission. Whatever the fine print of legal "agreements" supposedly consented to by entering a website, many felt concerned to assert and protect their "*rights*" to privacy.

Quasi-commons

Malls are usually private property. New urbanist planners may embed their rules in the legal foundation of government-issued building codes, or the covenants of a homeowners' association, while nominally "public" spaces are actually under private ownership. Websites exist in the evolving legal context of cyberspace. Such spaces can all be viewed as quasi-commons if they are formally under private ownership but share characteristics of commons, especially if this concerns not just how they are used, but how they are governed. Governance includes not just making and enforcing rules, but the constitutional level of rules for making rules: determining who has voice and authority to enact or adjust rules, how leaders and managers should be chosen, and how disputes should be resolved. In quasi-commons, users act not as co-owners of common property, but as stakeholders. However the characteristics of such spaces, and the ways in which users assert claims over their governance lead them to take on elements of commons, becoming quasi-commons.

Stakeholder Monitoring

One of the advantages of common property resource management has been that a community of users is much more able to monitor use, and detect abuse, than a private owner dependent on their own abilities or those of agents. Cheap, widespread videocameras change the cost of surveillance. This undercuts one of the advantages of common property management, to the extent that potential violators are deterred by the threat of detection. Other forms of automation and computerization also lower the relative costs of private management compared to collective action. Nanotechnology would reinforce these reductions in the costs of private monitoring. However to extent that norms are created and enforced not through explicit rule-making and formal enforcement, but rather mainly through interaction with other people, then other users still play a major role, even if the space is formally private or public property. Thus other users not only have a stake in being able to use and enjoy such spaces, but they play an important role in creating and maintaining rules, primarily informal, about how such spaces are used. Furthermore, and crucially for their influence in governance, without their presence the common space loses its value.

Coevolving Communities of Choice

Users choose whether or not to use quasi-commons. Rather than a community of fate, tied together by mutual interdependence, often within the confines of a small locality, and highly dependent on a single resource, these quasi-commons reflect the multiple, voluntary affiliations of an urbanized culture (though they may be physically located in the more dynamic and conducive conditions of "edge cities"). Exit by users is an important sanction, though its impact might mainly work through the concerns of merchants who want to maintain an attractive place to market their wares. Communities sharing commons and quasi-commons interact with each other, creating their mutual environment. Individuals and families usually maintain multiple affiliations. Principles are applied across different realms of experience. Standards emerge from debate and dialogue, and evolve over time. Such spaces, and their governance, are dynamic.

Constituting Governance in Quasi-commons

Bringing together these ideas, the influence of users as stakeholders in quasi-commons, affiliation by choice and coevolution among communities, implies that management of commons or quasi-commons is likely to be dominated not by direct democracy within small communities where everyone knows everyone else and works out problems through consensus. Instead, management is likely to mix modes of common property (such as homeowners associations), state regulation, and private enterprise, with stakeholder participation in governance of quasi-commons. Users' choices regarding entry and exit may have a bigger influence than voice on how quasi-commons are managed. The role of those sharing common spaces would come not as co-owners, but as stakeholders.

Crafting Commons for Millennia

Healthy, Wealthy and Wise

Nanotechnology promises material abundance. In the longer run it offers tools for curing illness, restoring health, and greatly prolonging life, the prospect of living for hundreds or even thousands of years. Further enhancement of computers and communications should make data, information and a wealth of knowledge widely available. Long life may bring a much longer time horizon for considering self-interest. On the other hand, accelerating change may further reduce time horizons for investments, emphasizing the importance of ephemeral commons, quickly created and perhaps quickly disappearing. In information commons, a reputation for honesty, creativity and other intangible assets may much more important than material or financial wealth. If opportunities are less and less constricted by material constraints, the most important challenges may concern choosing what to do from a widening array of choices, and creating new choices, including new commons. Fundamentally, nanotechnology may offer the time and the capabilities to explore the potentials for being human, or more than human. It is hard to know what directions these may take, but they may well return to the central concerns of religion and philosophy, and the mundane arts of living a satisfying life among family and friends.

Collective Enterprises

However it is almost inevitable that many of these activities will be collective, bringing together groups of people larger than families and smaller than nations to undertake joint enterprises. While governments and corporations are unlikely to wither away, they are likely to be accompanied by a greater variety of other institutions, already flourishing in the "Third Sector" of associations, clubs, trusts, foundations and other forms of organization for pursuing interests other than money and power. Knowledge about how to manage common property will continue to be useful, although as noted above, it may take forms distinct from more traditional institutions for managing common pool resources.

Visions

Nanotechnology is protean enough that it could serve many visions:

1. The development of nanotechnology may be strongly influenced by the underlying ideas of those who develop the technology, whether soft paths to green sustainability, hardedged aspirations for high tech expansion, pursuit of personal immortality, or various amalgams and alternatives. The evolution of computer technology was strongly influenced by those who pioneered personal computers, inspired by visions of individual empowerment, affordability and freedom from dependence on corporate resources. To the extent those involved in the development of nanotechnology share a similar individualistic or libertarian ethos, this may shape the way the technology is developed.
2. Edward Regis' exploration of nanotechnology explains the conceptual foundations, describes protagonists and outlines amazing potentials, but ends with a suggestion that much of the future might be captured by a vision of (an American) Sunday afternoon in the suburbs.
3. Some may strive to employ nanotechnology and other technologies to deal with what they see as major threats to the human species, whether preventing future ice ages, fending off asteroid impacts, or expanding humanity beyond a single vulnerable planetary ecosystem.

4. The Martian settlers in Kim Stanley Robinson's science-fiction trilogy add nanotechnology to their tool-kit, but focus their livelihoods on smallholder agriculture as part of ecopoesis, with plenty of leisure for pastimes such as hiking, sailing, and flying.
5. Richard Norgaard's book, *Development Betrayed* stands out from most critiques of development in its attempt to reexamine the conceptual roots of the modern project of universalist development, and to go beyond criticism to make some initial suggestions about an alternative prospect for a "patchwork discourse of coevolving communities." Just as the educational and emancipatory opportunities of computers and the internet have been incorporated in green visions, soft versions of nanotech may well find a welcome among those concerned with sustainable communities and ecological regeneration.
6. When asked what he thought of western civilization, Gandhi is said to have replied that "western civilization would be a good idea." Around the world, diverse cultures have aspirations that go far beyond their current achievements, and much that could be done to work towards realizing those aspirations. While current conceptions of nanotechnology are heavily framed by Western and particularly American ideas of progress, it is not necessarily restricted to the implementation to any specific cultural agenda, nor limited to the intellectual projects of the Enlightenment and Modernity.
7. The continued development of computer technology may make it possible to transfer human minds into a computers. If nanotechnology helps make such "uploading" possible, then the governance of cyberspaces would become even more crucial. Living in virtual realities at computer speeds would also make it possible to experience subjective millennia in relatively short periods of objective time.
8. Quite likely, future enterprises may take forms hard to envision now, but still face challenges of how to create and manage common property.

Institutional Architecture

Long life and prosperity may make it possible to explore an expanding diversity of ways to live, whether pastoral or high tech, earthbound or traveling to the stars. These various enterprises all face the challenges of organizing collective action in shared spaces, commons, whether in the material world or cyberspace. Long life could make it possible to undertake truly long-term projects. Managing coexistence among diverse communities may pose major challenges. Structuring incentives, assuring cooperation, become even more important in the design of institutions for the long term, including the ability to reconstitute institutions in the face of unforeseen challenges. Longer-term perspectives could stimulate greater interest in the art and science of institutional architectures for common property, the challenges of constituting commons.

Conclusions

Exploring a few of the many possible consequences of nanotechnology indicates how it might bring profound implications for the management of existing commons and the creation of new commons. Prediction is impossible, but scenarios show how it may be important to reconsider some of the assumptions that underlie much study and management of common property. Nanotechnology itself could be developed as an abundant common property information resource, a new commons, applying the principles behind open source computer software. Recycling at the elemental level, assembling and disassembling materials out of their constituent atoms, may promote resource abundance, converting one way flows from extraction, consumption and disposal into closed cycles, reusing and managing resource stocks.

If this furthers declining prices for commodities, then it may reduce incentives for private appropriation, and sometimes favour common property management. Stakeholders may become even more assertive in reconstituting governance of shared private and public spaces as quasi-commons, mixing modes of state, private and common property management. Lifetimes measured in centuries or millennia, are likely to encourage new projects, collective enterprises pursing goals distinct from private profit or state survival, some of which may well be governed as common property. Nanotechnology may make common property more common, not less, with commons, in physical space and cyberspace, increasingly important in enabling people to pursue their dreams. The study of common property should continue to offer useful intellectual tools that people may use to craft their futures. "The best way to predict the future is to invent it."

11

GENETICS AND NANOTECHNOLOGY

The paper opens with the question raised by Grundmann and Stehr, as to whether "knowledge policy" may include "the aim of limiting, directing into certain paths, or forbidding the application and further development of knowledge". It then explores this theme with reference to contemporary developments in *biotechnology* and *nanotechnology*, where the objective of knowledge is to enable us to create and modify at will biological entities (including humans and combined species known as "*chimeras*"), as well as self-assembling mechanical entities, ab initio through recombinant DNA techniques. A new category of risks is created by the promised technological applications of these forms of knowledge, called "*moral risks*", which threatens the ethical basis of human civilization; these are also "*catastrophic risks*", in that their negative and evil aspects are virtually unlimited. The paper asks whether our institutional structures, including international conventions, are robust enough to be able to contain such risks within acceptable limits; or alternatively whether these risks themselves should be regarded as unacceptable, a position which would impel us to seek to forbid individuals and nations from acquiring and disseminating the knowledge upon which those technologies are based.

INTRODUCTION: "EPPUR SI MUOVE"

At the conference "The Governance of Knowledge", Essen, Germany, September 5-7, 2001, Reiner Grundmann and Nico Stehr presented the background paper "*Policing Knowledge*: A New Political Field" which poses "the question of social surveillance and regulation of knowledge". They suggest that "knowledge policy" may include "the aim of limiting, directing into certain paths, or forbidding the application and further development of knowledge". If scientific knowledge is included here, this proposition will not be well received.

One of the great founding faiths of modern society is that of the infinite benefits of the liberation of the natural sciences from the intellectual and institutional shackles of dogma, including religion; its inspirational image is that of Galileo before the Inquisition, forced to recant publicly his belief about earth's movement in space, but unyielding in his mind and certain subjectively of his ultimate vindication. Anyone who seeks to challenge this faith is in for a rough ride. Are there forms of knowledge about nature (including a technological capacity to manipulate nature based on them), now envisioned as practical possibilities in foreseeable futures, of which it may be said that they are too dangerous for humanity to possess? Too dangerous, at least, in the hands of that radically imperfect humanity in and around us, including its all-too-delicate veneer of civilization, which now seems prepared to seek that knowledge? And if so, is it even conceivable that one could argue for their suppression on the grounds that, once realized they will inevitably be deployed, to ends so evil, running unhindered into the future, as to destroy the moral basis of civilization? I at least am not ready to answer these questions – although they are being raised by some in the academic community, especially with reference to biotechnology.

An editorial earlier this year in *New Scientist*, commenting on the inadvertent laboratory creation of a virulent engineered virus which could be used as a weapon in biological warfare, said: There's also the problem that many biologists choose to ignore biotechnology's threats.... John Steinbruner of the University of Maryland, College Park, has suggested setting up bodies to oversee areas of biological research. Such bodies could question or even stop research, or decide if results should be published. As Steinbruner is well aware, his proposal strikes at the heart of scientific openness and freedom. But leaving things as they are is not an option. Biotechnology is beginning to show an evil grin. Unless we wipe that smile from its face, we'll live to regret it.

Catastrophic risk in this sense can be defined as the possibility of harms to humans and other entities that call into question the future viability of existing animal species, including our own. Thus these are not only risks to the present generations of living animal species, but also to future (perhaps all future) generations of presently existing species. One well-known risk of this type is what has been called "*nuclear winter*", the threat of a pervasive environmental catastrophe that could follow a large-scale exchange of nuclear weapons between the United States and the former Soviet Union (now Russia), under the doctrine of "*mutually assured destruction.*" The hypothesis of environmental catastrophe was based on the expectation that the earth's atmosphere would become loaded with particulate matter, blocking much of the solar radiation reaching the earth's surface, perhaps for a period of years (such an event is thought to have occurred following the impact of massive asteroids colliding with the earth). In addition, of course, the huge doses of radiation emitted by these exploding weapons would have profound genetic consequences for plants and animals.

The Lords of Creation

Given the existing stockpiles of nuclear weapons, the risks associated with them still exist, although (in view of the political instability in Russia) it is difficult to know whether the probability now is greater or less than before. *But new catastrophic risks are on the horizon, and these have a fundamentally different character that may require very different institutional responses from us.* Their common characteristic, considered as basic and applied science and the technological applications made possible through them, is that they are all based on our latest understanding of biological systems through molecular biology.

More specifically, their common scientific basis is the capacity to characterize complete genomes and to manipulate them by means of recombinant DNA techniques (or to create DNA-like mechanical structures). The ultimate goal, already envisioned and set as an objective for research, is a knowledge of *genomics* so complete that living entities (and life-like mechanical entities) could be constructed, or alternatively deconstructed and then rebuilt and varied, *ab initio*. According to an article published in *Science* in 1999, researchers working with a microbial parasite sought to characterize and develop "an organism with a minimal genome, the smallest set of genes that confers survival and reproduction": But since each of the 300 genes found to be essential could have multiple functions (pleiotropism), investigators had no way of finding the degree of redundancy and whittling the genome down further. The next logical step: make a synthetic chromosome of just those genes to build a living cell from the ground up.

Considered in their human implications, regard these developments as giving rise to a new type of catastrophic risk, which here called "*moral risks*". Gradations of being (inorganic and organic matter, plants, insects, animals, humans) are and always have been a foundation-stone of humanity's ethical and religious systems. More particularly, "*self-consciousness*" has been regarded as the essential and distinguishing mark of a human being, uniquely; yet as illustrated in the following section we have, apparently even among some senior scientists, an inclination to experiment with "crossing" these dimensions of existence in an almost casual mood.

Short List of "Catastrophic Risks"

There are risks from the use of future bioengineered pathogens used as weapons or war or terrorism. A recent review in *Nature* listed the following possibilities:

1. Transferring genes for antibiotic resistance (e.g., to anthrax or plague, as Russian scientists have done) or pathogenicity (the toxin in botulinin, which could be transferred to *E. coli*), or simply mixing various traits of different pathogens, all of which is said to be "*child's play*" for molecular genetics today.
2. Through "*directed molecular evolution*", especially what is called "*DNA shuffling*", producing "*daughter genes*" by shattering genes and then recombining gene fragments in ways that change the natural evolutionary pathways of bacteria.
3. Creating "*synthetic*" pathogens, that is, "artificial" bacteria and viruses, by starting with a synthesized "*minimal genome*" which was capable of self-replication (a kind of empty shell), to which "*desired*" traits could be added at will.
4. Creating hybrids of related viral strains.

These possibilities multiply as scientists begin publishing the complete DNA sequences of well-known pathogens: "... [G]enomics efforts in laboratories around the world will deliver the complete sequence of more than 70 major bacterial, fungal, and parasitic pathogens of humans, animals and plants in the next year or two...." Scientists working in these areas point out that actually getting engineered viruses and bacteria to survive in the environment, and to be maximally useful as weapons of war and terrorism, would not be easy to do; moreover, defenses against them can be constructed. What we are faced with the advances in molecular genetics, therefore, is an increase in the risks of novel agents being used in these ways for nefarious purposes.

There are related risks from accidental or unintended consequences of genomics research, especially from the genetic engineering of viruses and bacteria, which could result from the escape into the environment of virulent new organisms, irrespective of whether these organisms were intended originally for "*beneficent*" or "*malevolent*" purposes.

There was a brief flurry of publicity earlier this year when Australian researchers announced that, in engineering the relatively harmless *mousepox virus* with a gene for the chemical interleukin 4, in an attempt to create a contraceptive vaccine for mice, they had accidentally made the virus exceptionally toxic: "The virus does not directly threaten humans. But splice the IL-4 gene into a human virus and you could create a potent weapon. Add the gene to a pig virus, say, and you could wreck a nation's food supply".

There are risks to the "*nature*" of humans and other animals from intended or unintended consequences of genetic manipulations that either introduce reproducible changes into an existing genome (e.g., human or animal germline gene therapy), thus modifying existing species, or create entirely new variant species. For illustration here, I will confine myself to the example of "*chimeras*", that is, combined entities made up of parts of the genome of two or more different species, including of course humans. Some molecular biologists apparently already have done casual experiments inserting human DNA into the eggs of other animals and growing the cell mass for a week or so; and there is much speculation as to what would happen if human and chimpanzee DNA were crossed, since chimps share over 98 % of human genes.

The DNA of all species now on earth is composed of the same four chemical bases, abbreviated A, T, C, G, arranged into two pairs (A/T, C/G), that make up the "*ladders*" on the double helix of DNA; different combinations of the base-pairs specify one of 20 amino acids, which combine to form various proteins. Some scientists are experimenting with adding more chemicals that would act as new bases, so that, for example, there would be six rather than four bases and perhaps three base-pairs. One of the scientists doing this work is Peter Schultz: "Schultz often says living things have only 20 amino acids because God rested on the seventh day. 'If He worked on Sunday,' he said, 'what would we look like?'" The self-comparison between Dr. Schultz and God is interesting, to say the least.

There have been widely-publicized discussions of certain unique risks to organic life, stemming from possibilities allegedly inherent in the development of robotics and nanotechnology, especially in a now-infamous paper by Bill Joy (April 2000), Chief Scientist at Sun

Microsystems and creator of the "*Java*" script. Joy wrote: The 21st-century technologies - genetics, nanotechnology, and robotics (GNR) - are so powerful that they can spawn whole new classes of accidents and abuses. Most dangerously, for the first time, these accidents and abuses are widely within the reach of individuals or small groups.... I think it is no exaggeration to say that we are on the cusp of the further perfection of extreme evil, an evil whose possibility spreads well beyond that which weapons of mass destruction bequeathed to the nation-states, on to a surprising and terrible empowerment of extreme individuals.

The link between nanotechnology and biotechnology is fascinating: Although the former works with intrinsically inert materials, it is seeking to turn them into a perfect analogue of a biological system. One of the leading Canadian scientists in this field, Dragon Petrovic, has explained the quest as follows: In the future, he predicts, technicians will teach individual molecules and atoms to assemble themselves into wires and sheets of impeccable purity and thinness.... [Imagine] instruments made of compounds that are self-assembled, atom by perfect atom - materials so pure that they could never snap apart or break under normal conditions.... "Imagine [Petrovic says] the linkage to telecom - can we get DNA molecules to self-assemble into perfect sheets and wires only an atom thick, and then send electrons and photons to stimulate the DNA to do things - start growing; stop growing; assemble into certain geometric shapes? It's analogous to what a structure like bone does in nature, where the brain is the electronic device and the nervous system transmits the information".

Bill Joy's essay already had explored the dark side possibly inherent in the quest for self-replicating nanotechnology machines; the internal quotation in the passage by Joy below is from a book by Eric Drexler, *Engines of Creation*: An immediate consequence of the Faustian bargain in obtaining the great power of nanotechnology is that we run a grave risk - the risk that we might destroy the biosphere on which all life depends. As Drexler explains: Tough omnivorous "*bacteria*" [created by nanotechnology] could outcompete real bacteria: They could spread like blowing pollen, replicate swiftly, and reduce the biosphere to dust in a matter of days.... Among the congnoscenti of nanotechnology, this threat has become known as the "*gray goo problem*".

The "*gray goo problem*" attracted so much attention that in England the Royal Society and the Royal Academy of Engineering commissioned a special expert report on it: "*Nanoscience* and *nanotechnologies*:

Opportunities and Uncertainties" (July 2004). This report contained a special appendix on the "problem", which, it suggested, represented a remote and dubious risk; but it also addressed some unique and quite relevant risks, associated with nanotechnologies, which will be a challenge for government regulatory regimes to come to grips with.

One important point must be emphasized here, namely, that what has been just described are (hypothetical) catastrophic "*downside risks*", that is, the potential for very great harms to be done through some future technologies that are already on the drawing-boards. For each of these developments there are both "*upside benefits*", resulting from future applications of these technologies that could bring substantial benefits to us, as well as the potential for "*protective*" technological innovations that could mitigate, offset, reduce, or even eliminate at least some of the downside risks. To take the example of the engineering of viruses as *bioweapons*: As a counter to this threat (and also just to reduce the debilitating effects of viral infections on population health), research is under way in molecular genetics to develop new antiviral drugs that can block the infectious action of any viruses at the cellular level (preventing receptor binding, cell penetration, replication, production of viral proteins, and so on).

Considered as a totality, however, what these conjoined prospects do is to continually "*raise the stakes*" in our technological game with nature, whereby the new sets of risks and benefits reflect both, and simultaneously, the potential for an upside of hitherto unattainable benefits and a downside of hitherto unimaginable horrors. As discussed in a later section, this entire prospect increases the challenge to our social institutions to manage our technological prowess so as to realize the benefits and avoid the harms, and likewise increases the risk that we will be unable to do so.

What is Different Today?

There are undoubtedly other types of catastrophic risks, but those introduced above are sufficient for purposes of discussion! My main point is that these newer risks are fundamentally different in character from the case of nuclear winter, and the difference has to do with the distribution of knowledge and technological capacity relevant to them (thus requiring a very different institutional response). The technologies giving rise to the nuclear winter risk are controlled by just two nation-states and are maintained (for the most part, and until now) under a thick blanket of military security and secrecy, although the smuggling of nuclear materials out of the former Soviet Union is cause for worry.

Both the essential theoretical knowledge, and the engineering capacity needed to turn that into weapons, is confined to a relatively small circle of experts and officials. Not so with the new technologies.

The catastrophic risk areas listed above stem from current research programs that are widely distributed around the world; moreover, the strongest drivers of them are private corporations, including the large pharmaceutical multinationals, acting with full encouragement, support, and incentives from national governments. Especially where the possible health benefits of genetic manipulations are concerned, the combined public-private interests are overwhelmingly supportive, driving the research ahead at an accelerating pace.

Governments especially are enthralled with the economic significance of these new technologies, are competing with each other under innovation agendas to capture major shares of the corporate investments, and are loathe to stop and think about unintended consequences.

All of the characteristics of the knowledge and applications in these areas mean that it is extremely difficult even to think about controlling either the process or the results. For one thing, the knowledge is widely distributed among individual scientists; for another, it is widely distributed among private actors (corporations) which have the option of moving their operations on a regular basis, seeking perhaps the least-regulatory-intensive national base on the globe. Third, the technologies themselves become increasingly "*simplified*" and thus easier to hide, if necessary; the genetics technologies, for example, can be carried out in small laboratories almost anywhere.

Sergei Popov, the Russian scientist who pioneered germ warfare research using recombinant DNA techniques, observed recently: "The whole technology becomes more and more available. It becomes easier and easier to create new biological entities, and they could be quite dangerous".

Fourth, oversight is inhibited by the lure of truly extraordinary economic and health benefits promised by the new knowledge and technologies. And fifth, just the astonishing pace of innovation itself today makes the prospect of control and regulation a challenge.

During the past year national governments have been scrambling to respond to just a few of the dimensions of these new risks. Most attention has been focused on human cloning, where a few rogue scientists have challenged authorities in various jurisdictions to "*try to stop us*", and laws prohibiting this technology are being passed rapidly. But this is a relatively crude technology, albeit one which excites public attention, and one wonders whether authorities will become complacent about their ability to control unacceptable technologies due to their experience with this case. (Meanwhile, there are increasing reports that many genetics scientists are "going underground", in the sense that they have stopped talking publicly about their research in progress for fear that public reactions will be hostile and will result in official steps to halt it.)

Among the scientists cited in this chapter, two have called for urgent action under the Biological and Toxic Weapons Convention (1975, hereafter BTWC), to provide explicitly for a global oversight effort over some of the new technologies and their applications described earlier. Unfortunately, and ironically in view of what was to happen only two months later, at a meeting of the parties in Australia in July 2001 the United States unexpectedly blocked the process of completing a protocol under the BTWC that would have made the Convention something other than a statement of good intentions, for in its present form it has no provisions for verification or compliance monitoring. The US government has been pressured by its biotechnology industry sector not to agree to a verification protocol, under which inspections of laboratories and other facilities by international teams of experts would be carried out in all the signatory countries, because industry fears that its intellectual property and commercial secrets could be compromised. At the time of writing other signatories were considering whether they should proceed to complete the adoption of the verification protocol without US support.

Unfortunately, we know international negotiation to be at the best of times a tedious and protracted process, and there is reason to believe that in this domain it could be fractious and unsuccessful. This is because all of the technologies described represent frontiers of industrial innovation in which great multinational corporations and the national governments which protect their interests (especially the United States) have significant investments; both corporations and governments would be loathe to see those investments and the immense payoffs expected from them jeopardized by an international control regime. A recent article co-authored by a molecular geneticist and a specialist in the international convention on biological weapons has called for an urgent new effort to strengthen verification under the 1975 Convention and to enlist the biomedical research community in an effort to strengthen deterrence against the uses of bioengineered organisms for war and terrorism.

Conclusion

Now is the time for intensive exploration of the theme of policing science and to ask the following types of questions:

- Can we characterize a set of new catastrophic risks, as defined here, related to the leading-edge technologies that are being developed?
- Do these new risks have an essential character that will make them difficult to control, because the knowledge and the technologies will be so widely diffused?
- Can these risks be confined to acceptable dimensions by the institutional means now at our disposal, including international conventions on prohibitions? If not, what new tools do we need, and how can we get them?
- Do professional associations of scientists working in these fields have special responsibilities to assist societies in controlling these risks, and if so, are those responsibilities now being discharged adequately?

What is at risk in this game, now, is the possibility that the tension between science and society will become both unmanageable for institutions and unbearable for individuals, in other words, that the destructive applications of our operational power finally will overwhelm the rest. This possibility arises out of the striking contrast between the pace of change in social and legal institutions (especially international agreements), on the one hand, and in new scientific and technological breakthroughs in the sciences, especially in genomics, including applications relevant to biowarfare and bioterrorism on the other.

In the first-mentioned the pace is painfully slow and progress often remains ineffective even after decades of negotiation, as in the case of the Convention on Biological and Toxic Weapons. The second proceeds at a frenetic and steadily-accelerating pace. To reduce the probability that change in the second will overwhelm our social and legal capacity to steer technological development away from the zone of catastrophic risks, it is necessary first to get agreement among influential social actors that this is, as described here, a momentous challenge which contemporary society cannot avoid.

The first practical test of our resolve in this regard, is whether influential scientists can be mobilized in the cause, scientists who will reaffirm the need for new oversight structures, to be erected both within the practice of science itself and also in the relation between science and society. Hegel made a remark, somewhere in his writings, to the effect that only the hand which inflicts a wound can heal it. The wound here is the rupture with the dominant pre-modern relation of humanity and nature, governed by value-laden categories of being, and its replacement by modern science's purely operational orientation to the totality of the natural world. I will not speculate here on what a healing of that rupture could mean now, at least, not in any "ontological" sense. But in a practical sense, as a matter of public policy, I think it is clear what is required - namely, that the practitioners of science join others in a program to try to bring our operational powers under the control and direction of social institutions that have universal validity, ones that correspond in sufficient measure with the common aspirations of humanity.

It is my contention that today's dominant institutions do not have such validity and that, as a result, everyone on earth is at risk of having these powers become instruments in an Armageddon waged to the bitter end by contending social, ethnic, national, and religious interests. What remains to be seen is whether the task as defined here can be widely recognized and grasped as such, while there is still time, and whether our scientific enterprise can be steered towards the shelter of a social compact having universal validity.

If it turns out that despite our best efforts this cannot be done, there will arise a set of other questions that, for now at least, are too abhorrent for many even to consider. These questions have to do with the possibility that, taking both "normal" human passions and human institutional failings into consideration, there may be forms of knowledge that, as a practical matter, are too dangerous for us to possess, and that our only choice is to renounce and suppress such knowledge or suffer the consequences. In mentioning them we go to the heart of the fateful compact between science and society that has set the course for the development of modern society from the seventeenth century onwards, under the program known as the domination of nature. It is likely that contemporary society is not ready to deal with them, at least, not yet.

12

NANOMATERIALS

The union of distinct scientific disciplines is revealing the leading edge of Nanotechnology. Fifteen to twenty years ago, the inter-disciplinary activity of geneticists, biologists, immunologists and organic chemists created a more diverse toolbox now known as life science. *Bioconjugates* were created to help us move from outside the cell to the inside. Enabling technologies brought about the ability to create, identify and specifically manipulate genetic maps to engineer designer proteins. In parallel, physicists, chemists, polymer chemists and engineers were creating the foundation for the small world of nano materials science. *Fullerenes*, *carbon nanotubes* and *atomic force microscopy* were in their infancy.

In less than a decade, materials science and life science together are unraveling the mysteries of controlling, on a molecular level, the structure of matter. Particles, complexes, tubes, coatings, active surfaces and devices are being explored on the nanoscale. Assembly of nature's building blocks (e.g. carbon, nucleic acids, lipids and peptides) along with the combination of different materials (e.g. CdSe/ZnS, Au, Ag, Si(n)OH(n), light harvesting dendrimers and thin films) are leading to insightful understanding and the creation of new scientific tools. Chemists and physicists have been manipulating matter on the molecular level for centuries. Some say this is nothing new.

When one looks at the absolute elegance of the nanometer scale biological system, however, one is compelled to create, understand, manipulate and control systems with equal elegance. This high level of activity and promise has attracted private and government funding with considerable economic impact and growth. Since the launch of the National Nanotechnology Initiative in 2000, there have been hundreds of start-up companies emerging in the market. At universities, there are increased investments in nanotechnology programs and facilities. Northwestern University's Institute for Nanotechnology is a direct result of the high level of activity and pioneering work of Prof. Chad Mirkin.

As a result of massive grant support and continued focus from the university, the Center for Nanofabrication and Molecular Self Assembly building has been constructed. Similarly, The Molecular Foundry, a Department of Energy Nanoscale Science Research Center, is under construction at Berkeley in California. The research center is focused on the dissemination of nanoscale techniques and methods to enable scientists to delve into nano research. The Molecular Foundry is the direct result of Prof. Paul Alivasatos' discoveries and benchmark work. We see an increase in programs, facilities, career opportunities and educational outreach. It is a massive explosion of activity in study on a very small scale. Science is beginning to set free nanotechnology. Soon the only limit will be the imagination.

Nanoparticle-catalyzed Growth of Semiconductor Nanowire

Answers from the Past

Over the past twenty years, homogeneous nucleation in solution has proven to be an effective method for the synthesis of both *metallic* and *semiconductor nanocrystals*. Nearly *isotropic*, *pseudospherical* morphologies are typically produced. Other nanocrystalline morphologies are often desirable, and homogeneous nucleation has been adapted for the synthesis of rod-shaped morphologies. However, the growth of nanowires by homogeneous nucleation is apparently limited to materials having particularly favourable, highly anisotropic crystal structures. How then might nanowire morphologies be generally prepared?

The answer was uncovered by Wagner and Ellis in 1964, who discovered that micrometer-scale silicon whiskers (wires) could be grown from gold-droplet catalysts under *Chemical Vapour Deposition* (CVD) conditions at about 1000 °C. The process was named the "Vapour-Liquid-Solid" (VLS) mechanism after the three phases involved. VLS growth was extended to micrometer-scale whiskers of many inorganic materials and intensively studied for over a decade, before fading into relative obscurity.

From VLS to SLS

In 1995, my group reported that III-V semiconductor nanowires could be grown in solution from indium nanoparticle catalysts by a process analogous to the VLS mechanism, and by analogy we named it the "*Solution- Liquid-Solid*" (SLS) mechanism. Shortly thereafter, Lieber and coworkers described laser-ablation adaptations of the VLS mechanism that afforded silicon and a wide range of other semiconductor nanowires (*Laser-Assisted Catalyzed Growth*, LCG). Many others made seminal contributions to nanowire growth by VLS adaptations, most notably Korgel (*Supercritical Fluid-Liquid- Solid, SFLS, mechanism*) and Yang. We now routinely prepare soluble, III-V and II-VI semiconductor nanowires with controlled diameters in the strong-confinement regime of about 3–20 nm by bismuth-nanoparticle-catalyzed SLS growth. The nanoparticle-catalyzed VLS mechanism and its solution-phase variants have emerged as the popular, widely practiced, general methods for the synthesis of semiconductor nanowires. Gold nanoparticles are presently by far the most commonly employed catalysts.

Quantum Wires to Nanophotonics

The advantages of nanoparticle-catalyzed nanowire growth include its general applicability to a wide variety of materials, the diameter control afforded, the uniformity of the wires (lack of significant diameter fluctuations) and their oriented, near-single crystallinity. The surface passivation, solubility and length of the wires may also be systematically varied. Small-diameter "*quantum*" wires are ideal specimens for fundamental studies of two-dimensional (2D) quantum-confinement phenomena and for property comparisons to 3D-confined quantum dots, 1D-confined quantum wells and anisotropically 3D-confined quantum rods. Potential applications of semiconductor nanowires in nanophotonics and lasing, nanoelectronics, solar-energy conversion and chemical detection are under active development. Exciting progress and advances in the semiconductor-nanowire field, enabled by the emergence of nanoparticle-catalyzed growth, are anticipated in the immediate future.

Nanomaterials in Organic Photovoltaic Devices

The ability to create high-efficiency solar cells is a key strategy to meeting growing world energy needs. Nanotechnology is currently enabling the production of high-efficiency organic photovoltaics (OPVs) to help meet this challenge. Organic photovoltaics are nanostructured thin films composed of layers of semiconducting organic materials (polymers or oligomers) that absorb photons from the solar spectrum. These devices will revolutionize solar energy harvesting, because they can be manufactured via solution-based methods, such as ink-jet or screen printing, enabling rapid mass-production and driving down cost.

OPVs currently lag behind their "inorganic" counterparts because of low solar energy conversion efficiencies (approximately 1-3%). Several research groups are addressing conversion efficiency by employing a combination of nanomaterials and unique nanoscale architectures. These hybrid organic-inorganic photovoltaics consist of light-absorbing polymers in contact with semiconductor nanocrystals, fullerenes or nanostructured metals. The nanomaterials affect electro-optical properties of the conducting polymer, which include assisting in absorption of red and near-IR photons, a significant portion of the solar spectrum. Examples of OPVs designs employing nanomaterials include:

Polymer-fullerene Heterojunctions

Cells where a chemically modified fullerene (C_{60}) layer, acting as electron acceptors, is in close physical contact with a polymeric organic electron donor (MDMO-PPV catalog number 546461, or poly-(3-hexylthiophene), P3HT, catalog number 445703 and 510823). This contact improves efficiency by allowing charge transfer to take place at the sub 10-nanometer scale, on the order of the diffusion length of an exciton generated from organic semiconductors. The most recent cells exhibit conversion efficiencies of ~5%.

Organic-nanocrytsal Solar Cells

Blends of semiconducting polymers and semiconducting quantum dots or nanorods (CdSe or CdTe) are mixed in a manner similar to the polymer-fullerene blends. The polymers are modified to give rise to chemical bonding between the nanocrystal and polymer. The nanocrystals can be tailored to a wide variety of optical band gaps, which depend on the size of the nanocrystal (or the diameter of the nanorod).

Dye-sensitized Cells

These cells employ complex dye molecules attached to the surface of nanostructured oxides like titanium(IV) oxide (TiO_2) or niobium(V) oxide (Nb_2O_5). The dyes exhibit broad light-absorption profiles and rapid photoinduced charge transfer of electrons to the nanocrystals. These cells show solar conversion efficiencies of ~4%.

Tandem Cells

The tandem cell acts as a 2-1 cell, harvesting photons from the complete visible spectrum. These cells employ layers of C_{60} as a strong blue light absorber and copper-phthalocyanine (CuPc) as a red-yellow absorber. Nanosized silver particles act as a charge conduit between the cells but do not absorb photons traveling through the cell because of their nanosized dimensions. These cells have achieved conversion efficiencies of ~6%.

Significant challenges exist to achieving OPV devices that can be mass-produced. Nanotechnology will assist in meeting the technical challenges of this rapidly evolving field.

COATINGS

Nanopowders and nanoparticle dispersions have seen increasing applications in coatings. Due to their small size, very even coating can be achieved by painting nanoparticle dispersions onto a surface and baking off the residual solvent.

Optically Transparent Conductive Coatings

Indium tin oxide (ITO) and *antimony tin oxide* (ATO) are well known, optically transparent, electrically conductive materials. Nanoparticles of these materials can be painted on surfaces such as interactive touch screens to create a conductive, transparent screen without relying on expensive sputtering techniques. In addition, ITO and ATO can be used as an antistatic coating, utilizing their inherent conductivity to dissipate static charge.

Optically Transparent Abrasion-resistant Coatings

Nanoscale aluminum oxide and titanium oxide are optically transparent and greatly increase the abrasion resistance of traditional coatings. Titanium oxide is of particular interest in many optical applications, since it is highly reflective for most ultraviolet radiation. Zinc oxide and rare-earth oxides are also UV-reflective, but optically transparent and are therefore effective in protecting surfaces from degradation brought about by exposure to UV radiation.

CARBON NANOTUBE-BASED BIOSENSORS

Bioconjugates

Researchers from the University of Illinois at Urbana-Champaign have developed near-infrared optical biosensors based on single-walled carbon nanotubes, which modulate their fluorescence emission in response to specific biomolecules. The viability of sensor techniques was demonstrated by creating a *single-walled carbon nanotube* (SWNT) *enzyme bio-conjugate* that detects glucose concentrations.

Carbon nanotubes fluoresce in a region of the near-infrared, where human tissue and biological fluids are particularly transparent to their emission. The sensor could be implanted into tissue, excited with a near-infrared light source, and provide real-time, continuous sensing of blood glucose level by fluorescence response. As shown in figure a schematic mechanism of the nanotube sensor. Hydrogen peroxide is produced when glucose reacts with the enzyme, which quickly transforms ferricyanide to modulate near-infrared fluorescence characteristics of the nanotube.

Future Opportunities

The important aspect of this technology is that the technique can be extended to many other chemical systems. New types of non-covalent functionalization are developed, creating opportunities for nanoparticle sensors that operate in strong absorbing media of relevance to medicine or biology.

Layer-by-Layer Nanoassembly for Drug Delivery

Researchers from Louisiana Tech University were among the pioneers of one of the novel nanotechnology methods: *layer-by-layer* (LbL) nanoassembly by alternate adsorption of oppositely charged polyelectrolytes, nanoparticles and proteins. With this technique, we can assemble ultrathin multilayers with nanometer precision and pre-determined composition across the film, and make nanocapsules from LbL films. We are using such nanocapsules for targeted drug delivery, biocompatible nanocoating and pulp microfiber processing. For this nanoarchitecture, we use nanoblocks, such as nanoparticles. In the development of our research area, we successfully used nanoparticles, such as different diameter and surface-charged gold nanoparticles, silica, nanoclay-montmorillonite, alumina, titanium dioxide and other nanoparticles. Also, we investigated linear polyelectrolytes of different types and molecular weight (especially natural polyelectrolytes), which we used as electrostatic glue to assemble nanoparticle and protein arrays. As any architect, we nanoarchitects need a wide palette of nanoblocks with new properties and dimensions, ideally, monodispersed, stable in solution, charged nanoparticles of noble metals, metal oxides and ceramics with diameters of 5, 10, 20, 50 and 100 nm.

Antibacterial Applications

Zinc oxide is an effective antibacterial and anti-odor agent. It has been used in deodorants, dental cleansers and diaper creams. The increased ease in dispersibility, optical transparence and smoothness make zinc oxide *nanopowder* an attractive antibacterial ingredient in many heath care products. Copper oxide nanopowder has also been proposed as an anti-microbial preservative for wood or food products.

NANOMATERIALS IN FUEL CELLS

Over the last two decades, general interest and research in fuel cells has increased, because they have the potential to be more energy efficient than conventional power generation methods. During this time period, researchers have begun using nanomaterials in the catalyst layer of fuel cell electrodes for a variety of reasons, including: increasing the active surface area of the anode and cathode catalyst, increasing the catalytic rate of oxidation or reduction and minimizing the weight of platinum and other precious metals in the fuel cell.

The current generated at an electrode is proportional to the active surface of catalyst on the electrode surface, so higher power density fuel cells can be formed from nanomaterials, because nanomaterials have a higher surface area to volume ratio. Researchers have also shown that the electrocatalytic properties of the materials are sensitive to particle size, so increased catalytic activity can be observed for nanoparticles and nanomaterials.

However, the most important goal has been to decrease the weight of platinum and other precious metals in the catalyst layer of the fuel cell, so that the fuel cell can be cost-effective. This has been the main limitation to the widespread use of fuel cells. Researchers have employed carbon nanomaterials as supports for dispersions of platinum nanomaterials. This allows for a decrease in the weight of platinum needed to produce the same surface area of active platinum catalyst. The nanomaterials could be carbon foams containing nanopores, different types of nanotubes or even single-walled nanohorns. All of these materials can act as a support and a conductor for platinum nanomaterials, making strides toward cost-effective fuel cell catalysts.

NANOMATERIALS IN OTHER ENERGY APPLICATIONS

Environmentally Friendly Energy Sources

Devising schemes to meet the world's growing energy demands while simultaneously reducing green house emissions and other pollutants, has become one of the major challenges facing materials scientists. Nanomaterials promise to help solve many of the problems associated with new and emerging energy technologies.

Fuel cells

Solid oxide fuel cells (SOFCs) offer the advantage over other fuel cell designs in that they do not require expensive, precious metal catalysts and can operate effectively without extensive purification of fuel sources. The activity of doped rare-earth oxide electrodes such as yttrium stabilized zirconia (YSZ) is directly related to their surface areas. Nanoparticles exhibit the high surface required for developing SOFC technologies.

Cleaner emissions

Catalytic converters on vehicles around the world have significantly reduced the amount of automotive pollution over the last three decades. These devices require large amounts of expensive metals such as platinum, palladium and rhodium. Doped rare-earth metal oxides offer the promise of increased catalytic activity without the heavy reliance on precious metals. In addition, the increased efficiency of the next-generation catalytic converters will result in cleaner emission of existing internal combustion and diesel engines.

NANOTUBE-POLYMER COMPOSITES FOR ULTRA STRONG MATERIALS

Exceptional Strength

Exceptional mechanical properties of *single-walled carbon nanotubes* (SWNT) have prompted intensive studies of SWNT-polymer composites. However, the composites made with nanotubes are still holding a substantial reserve of improvement of mechanical properties. The problem is that pristine SWNTs have very poor solubility in polymers, which leads to phase segregation of composites. Severe structural inhomogeneities result in the premature failure of the hybrid SWNT-polymer materials.

The connectivity with the polymer matrix and uniform distribution within the matrix are essential structural requirements for strong SWNT composites. This problem is being solved by several approaches. First, by using coatings from surfactants and polymers, such as sodium dodecyl sulfate or poly(styrenesulfonate). This enables formation of better dispersions in traditional solvents including water. Polymeric dispersion agents are strongly preferred for the composite preparation because of (a) tighter bonding with the graphene surface, (b) miscibility with polymer matrixes of composites and (c) substantially smaller concentration necessary for the preparation of SWNT dispersions. Among polymers, different poly(vinyl alcohols) work best as a host matrix for SWNTs, providing the composites with high tensile strength and excellent Young's modulus.

Looking Ahead

An additional resource of SWNTs for improving the mechanical properties of composites is nanotube orientation. Virtually complete alignment of nanotubes can be achieved in SWNT composite fibers. These composites display substantially better mechanical properties than any other SWNT-polymer hybrid. If one can find a simple and controllable method to produce not only fibers but also bulk materials and coatings with nanotubes oriented in a desirable direction, new technological vistas can be opened for various composites. Breakthroughs in this area can come both from the studies of fundamental properties of SWNTs and from development of methods for composite processing. In the past, magnetic field alignment with exceptionally powerful electromagnets and alignment in the flow was used for this purpose.

Metal Matrix Nanocomposites for Structural Applications

Improved Properties

Metal matrix composites (MMCs) such as continuous carbon or boron fiber reinforced aluminum and magnesium, and silicon carbide reinforced aluminum have been used for aerospace applications due to their lightweight and tailorable properties. There is much interest in producing metal matrix nanocomposites that incorporate nanoparticles and nanotubes for structural applications, as these materials exhibit even greater improvements in their physical, mechanical and tribological properties as compared to composites with micron-sized reinforcements. The incorporation of carbon nanotubes in particular, which have much higher strength, stiffness, and electrical conductivity as compared to metals, can significantly increase these properties of metal matrix composites. Nanocomposites are being explored for structural applications in the defense, aerospace and automotive sectors.

Low Cost Solutions

Concurrent with the interest in producing novel nanocomposite materials is the need to develop low cost means to produce these materials. Most of the prior work in synthesizing nanocomposites involves the use of powder metallurgy techniques, which are not only high cost, but also result in the presence of porosity and contamination. Solidification processing methods, such as stir mixing, squeeze casting and pressure infiltration are advantageous over other processes in rapidly and inexpensively producing large and complex near-net shape components, however, this area remains relatively unexplored in the synthesis of nanocomposites. Stir mixing techniques, widely utilized to mix micron size particles in metallic melts, have recently been modified for dispersing small volume percentages of nanosize reinforcement particles in metallic matrices. Although there are some difficulties in mixing nanosize particles in metallic melts resulting from their tendency to agglomerate, a research team in Japan has published research on dispersing nanosize particles in aluminum alloys using a stir mixing technique. Researchers at the Polish Academy of Science have recently demonstrated the incorporation of greater than 80 volume percent nanoparticles in metals using high-pressure infiltration with pressures in the GPa range. Composites produced by this method possess the unique properties of nanosize metallic grains.

Recently, metal matrix nanocomposites were synthesized at the University of Wisconsin, Milwaukee using aluminum alloy A206 and nanoparticles of alumina (Al_2O_3). TEM samples of the cast Al-A206/Al_2O_3 clearly show nanoparticles present within the metal matrix. SAD patterns show the pattern of the matrix as well as the nanoparticles. EDX indicates that the grains are composed of aluminum, which contains nanosize alumina particles. The distribution of particles throughout the grains of the matrix with an absence of large concentrations at the grain boundaries suggests wetting of the alumina by the liquid metal. In this case the nanoparticles did not appear to act as nucleation sites for nanosized grains.

Clusters and Nanoparticles

Particles with a size between 1 and 100 nm are normally regàrded as nanomaterials. Figure 12.1 shows the size of nanoparticles in comparison with other small particles. In general, nanomaterials may have globular, plate-like, rod-like or more complex geometries. Near-spherical particles which are smaller than 10 nm are typically called clusters. The number of atoms in a cluster increases greatly with its

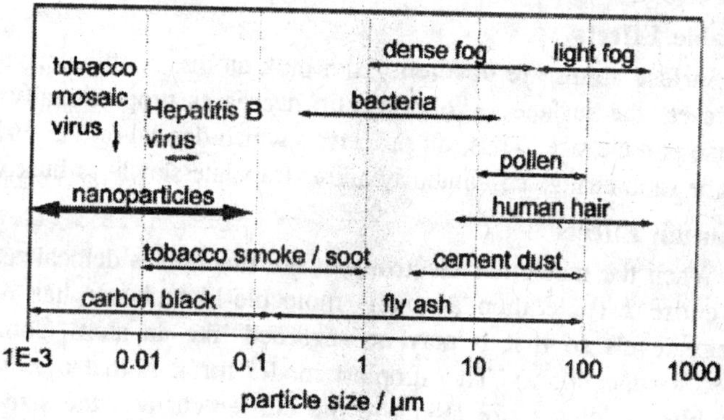

Fig. 12.1. Typical size of small particles.

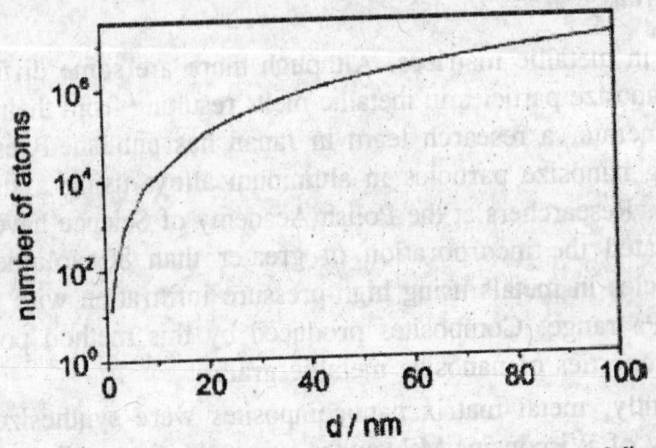

Fig. 12.2. Number of sodium atoms in a spherical cluster of diameter d.

diameter, demonstrated in Figure 12.2 for sodium clusters. At 1 nm diameter there are 13 atoms in a cluster and at 100 nm diameter the cluster can accommodate more than 10^7 atoms. Clusters may have a symmetrical structure which is, however, often different in symmetry from that of the bulk. They may also have an irregular or amorphous shape. As the number of atoms in a cluster increases, there is a critical size above which a particular bond geometry that is characteristic of the extended (bulk) solid is energetically preferred so that the structure switches to that of the bulk.

It is below a dimension of 100 nm where properties such as melting point, colour (i.e. band gap and wavelength of optical transitions), ionisation potential, hardness, catalytic activity and selectivity, or magnetic properties such as coercivity, permeability and saturation magnetisation, which we are used to thinking of as constant, vary with size. We basically distinguish two types of variations as a function of size:

Scalable Effects

Surface atoms are different from bulk atoms. As the particle size increases, the surface-to-volume ratio decreases proportionally to the inverse particle size. Thus, all properties which depend on the surfaceto-volume ratio change continuously and extrapolate slowly to bulk values.

Quantum Effects

When the molecular electronic wave function is delocalized over the entire particle then a small, molecule-like cluster has discrete energy levels so that it may be regarded like an atom (sometimes called a super atom). The simplest model for it is that of a particle in a box. Adding more atoms to the cluster changes the size of the box continuously so that the energy levels close up to some extent. More importantly, adding more atoms means adding more valence electrons to the system. Thus, whenever a shell of sometimes multiple degenerate energy levels is .filled the next electron has to be accommodated in the next shell of higher energy. The situation is analogous to the evolution of properties with increasing atomic number in the periodic table. Filled shells represent a particularly stable configuration. Properties such as ionisation potential and electron affinity are well known to display a discontinuous behaviour as one moves along the periodic table. For clusters consisting of atoms with strongly overlapping atomic orbitals, i.e. for metals and semiconductors, the situation is analogous.

Quantum effects are more pronounced for small clusters and often superimposed on a smoothly varying background of a scalable effect. Clusters are interesting intermediates between single atoms and bulk matter and represent a natural laboratory to 'see both ends from the middle'.

Feynman's Vision

On 29 December 1959, at the annual meeting of the American Physical Society, Richard Feynman addressed the audience with his visionary and by now historical and legendary lecture under the title - There is Plenty of Room at the Bottom: Invitation to Enter a New Field of Physics. With this talk on the problem of manipulating things on a small scale, Feynman opened the field of nanotechnology. Today, more than four decades later, the field is finally seen to really take off. It is amazing how closely some of the key developments follow Feynman's vision. Why cannot we write the entire 24 volumes of the Encyclopaedia Britannica on the head of a pin? he asked. We know how successful information technology has been in its work towards this goal and we should be aware of how much this has influenced our lives. Make the electron microscope hundred times better, Feynman said. The development of the atomic force microscope was one of the milestones on the way not only to observe but also to manipulate in atomic dimensions. Amazingly, Rohrer and Binnig achieved this goal with their cantilever-based instrument in a single step, rather than by a hierarchy of smaller and smaller robots - training an ant to train a mite - as Feynman suggested. In 1986, the two scientists were honoured for their achievement with the Nobel Prize in Physics. It is quite obvious today that the invention of the scanning tunnelling microscope finally triggered the boom in nanotechnology to which the direct observation of very small scale structures down to individual atoms is essential. Obviously, seeing things directly is more convincing for vision-based beings than just having measurements which are in agreement with a model.

Some of the inspiration came from biology. Feynman understood that information is stored on a molecular level in biology, that cells manufacture substances and operate on a small scale, that the human brain is a wonderful and efficient miniaturized computer. Consider the possibility that we too can make a thing very small which does what we want, an object that manoeuvres at that level, he suggested. While his talk was primarily technology oriented, he knew that physics, chemistry, biology and engineering are all relevant and must all be involved. He realized that making things smaller was not just a technological problem of scaling down. He saw that certain things changed principally. Magnetism, for example, is a cooperative phenomenon and involves domains which cannot be reduced down to an atomic size. Atoms differ from the bulk in their quantum nature. Most significantly, Feynman predicted that when we have some control of the arrangement of things on a small scale we will get an enormously greater range of possible properties that substances can have. It is exactly the arrangement of things on a small scale which is the foundation for all the excitement about nanomaterials and for the success of modern materials science.

INDEX

A

A dwarf, 2
Ab initio, 247
Adriamycin, 198, 202
Aerogels, 56
Agoraphobia, 189
Airbus airliner, 82
Akamba, 98
Akathisia, 187
Alcmaeon of Croton, 106
Alcoholism, 186
Alexander, 109
Alexandria, 109, 110
Alzheimer's disease, 188
Amateur, 190
Ambroise Pare, 114
American Physical Society, 59
Amoeba proteus, 165
Amorphous, 54
Analogously, 69
Anaphylaxis, 187
Androphobia, 189
Anemic, 198
Angioedema, 187
Anthracosis, 103
Antibiotics, 134
Antimony tin oxide, 260
Antiseptic, 129
Antitrust, 82
Antitrust system, 82
Antitumour, 199
Apocalyptic nightmares, 58
Applied research, 73
Arachnophobia, 189
Aristotelian, 109
Aristotle, 109
Artificial cells, 174
Artificial vitamins, 232
Asclepiades, 110
Asclepius, 105
Assemblers, 230
Astounding Science Fiction, 170
Astrology, 186
Atom-by-atom, 211
Atomic force microscope, 59
Atomic force microscopy, 256
Atomic scale, 35

B

Bacillaria paradoxa, 192
Bacteria, 8, 250
Baizer, 56
Barbiturates, 134

Basic research, 73
Belonephobia, 189
Beneficent, 249
Benzodiazepines, 134
Bioconjugates, 199, 256
Biomineralization, 56
Biotechnology, 245
Bioweapons, 251
Black-goo, 64
Bleomycin, 198
Book of wounds, 104
Bottom up, 7, 45
Bottom up technique, 47
Bottom-up fabrication, 60
Bottom-up route, 226
Buckminsterfullerene, 6
Bugs, 228
Bulimia, 186

C

Cachexia, 195
Carbon magnetic nanoparticles, 203
Carbon nanotubes, 11, 256
Carcinogenesis, 197
Caretakers, 196
Catastrophic risk, 245, 246
Cell cycle, 196
Charles II, 116
Charles Turner Thackrah, 117
Chemical Vapour Deposition, 257
Chemotherapeutic, 201
Chemotherapy, 198
Child's play, 248
Chimeras, 245, 249
Chirurgia, 107
Chlorpromazine, 134
Chocoholics, 186
Christian, 112
Chromophobia, 189
Clanking replicator, 5
Coffee, 186
Cola drinkers, 186
Colder, 29
Colloidal quantum, 29
Colloids, 55
Colon, 195
Condensed matter, 27
Corpus, 106
Cremophor, 204
Cro-Magnons, 97
Cross-capacities, 52
Cryonics community, 62
Crystalline, 54

D

Dacarbazine, 198
Daughter genes, 248
De jour, 212
Deficit model, 12
Dense medium plasma, 203
Department of Defense, 91
Department of Energy, 76, 92
Desired, 248
Destructive nanomachines, 64
Development Betrayed, 243
Development Principles, 68
Developmental constraints, 193
Diffusion limited aggregation, 192
Digital music revolution, 86
Digital versatile disks, 54
Dipsomania, 186
Directed molecular evolution, 248
Disease idealism, 143
Dispersion polymerization, 203
DNA shuffling, 248
Domination, 186
Downside risks, 251
Drexler vision, 11
Drexlerian dystopia, 8
Drug deliverers, 199
Dynamic random access memory, 54

E

E. coli, 146, 248
Egypt, 106
Electro migration, 52
Empedocles, 106
Emulsion polymerization, 203
End-stage renal disease, 217
Endocytosis, 198
Engines of Creation, 5, 33, 34, 39, 41, 42, 162, 173, 227, 250
Environmental Protection Agency, 76
Enzyme bio-conjugate, 260
Epistemic community, 214
Erasistratus, 109
Ethical problems, 66
Evipan, 134
Ex vivo, 100
Externalities, 237
Extreme nanotechnology, 4
Extreme ultraviolet light technique, 47

F

Federal Drug Administration, 76
Fiction, 43
Fistula clubs, 117
Flushes away, 52
Focused ion beam technique, 47
Folliculalgia, 187
Foodstuffs, 12
Foresight Guidelines on Molecular Nanotechnology, 68
Foresighting, 213
Friedrich Engels, 117
Fullerenes, 6, 256

G

Galen, 111
Galen of Pergamum, 110
Galenic medicine, 112
Gatekeepers, 196
General Electric, 93
Genetically modified, 12
Genetically modified organisms, 20
Genomics, 247
Giant magnetoresistance, 30, 53
Gift culture, 229
Gluttony, 186
GM crops, 12
GM position, 12
Gold rush, 77
Golden Rule, 74
Granularity, 52
Gravidophilia, 186
Gray goo, 5, 19, 64, 232
Gray goo problem, 250
Graywater, 235
Green, 8, 238
Green goo, 5
Grown, 15

H

Helicobacter pylori, 188
Hemophobia, 189
Herophilus, 109
Hertfordshire, 115
Hippocrates, 106
Hippocratic, 112
Hippocratic corpus, 106
Hippocratic medicine, 106
Holism, 152
Homo sapiens, 97
Honorius III, 113
Humors, 107
Hydrophobic, 55
Hyperthermia, 57, 199
Hypochondria, 186

I

IL-24, 205
Imagining the Future, 42
In cyto, 146, 150
In situ, 49, 170

In vacuo, 71
In vitro, 137
In vivo, 100, 146, 149, 169, 179, 180, 183
Indium tin oxide, 260
Indoor air, 236
Industrial policy, 79
Infinitely, 193
Initiation, 197
Institutions, 70
Intel, 21
Internet revolution, 86
Irrigation society, 86
Isotropic, 257

J

Java, 250
Jewish, 112

K

Karl Marx's, 117
Kevlar sheets, 11
King Asa, 112
Kleptomania, 186

L

Ladders, 249
Laser-Assisted Catalyzed Growth, 258
Law system, 81
Laws, 70
Layer-by-layer, 48, 261
Librium, 134
Light-emitting devices, 48
Listeria monocytogenes, 181
Lithium carbonate, 134

M

Machine-phase nanotechnology, 89
Machines, 16
Machines Bull, 79
Magnetic fluid hyperthermia, 57
Magnetic random access memory, 53
Malevolent, 249
Malignant conversion, 197
Map, 59
Martial, 110
Masai, 98
Material sciences, 49
Mda-7, 205
MDR1, 206
Medical nanomachines, 62
Mere, 211
Meta-materials, 30
Metal matrix composites, 264
Metallic, 257
Metallic nanopowders, 11
Metastasize, 197
Methodology, 58
Microelectromechanical systems, 4
Microprocessor, 1
Miniaturization, 60
Minimal genome, 248
Mobilophilia, 186
Molecular manufacturing techniques, 62
Monoculture, 82
Moore's first law, 45
Moore's Law, 89
Moore's second law, 45
Moral risks, 245, 248
Mousepox virus, 249
Mustine, 197, 198
Mutual capture, 78
Mutually assured destruction, 247

N

Nanno-biology, 28
Nano, 2, 44
Nano device, 76
Nanocameras, 63
Nanocrystals, 29
Nanodimension, 60

Nanoelectronic, 50, 54, 62
Nanogram, 2
Nanomachine, 76
Nanomaterials, 89
Nanomedicine, 99, 100, 174
Nanomicrophones, 63
Nanoparticles, 6, 9, 20, 29, 64, 89, 198, 207
Nanoparticulate soot, 9
Nanopowder, 261
Nanos, 2
Nanoscale, 63
Nanoscience, 2, 45, 250
Nanoscience, 89
Nanosensors, 63
Nanotechnological assembler, 70
Nanotechnologist, 93
Nanotechnology, 1, 2, 3, 10, 30, 44, 45, 56, 58, 60, 77, 88, 226, 245, 250
Nanotechnology revolution, 11, 86
Nanotoxicity, 20
Nanotubes, 29
Nanoweaponry, 63
Nanowires, 29
Napoleon, 119
National champion, 80
National Institute of Standards and Technology, 92
National Institutes for Health, 72, 92
National Nanotechnology Initiative, 88
National Science Foundation, 72, 88
Natural Capital, 238
Nature, 26, 33, 40, 248, 249
Nature Nanotechnology, 24
Neanderthal, 97
Near-infrared, 200
Neolithic farmer, 8
Neutropenic, 198
New nanoscale phenomena, 27
New Scientist, 246
Noctiphobia, 189
Non plus ultra, 63
Normative analysis, 58
Nuclear winter, 247
Numerical control, 92
Nymphomania, 186

O

Office of Munitions Control, 76
On Medicine, 111
Open access, 234
Ophidiophobia, 189
Optical proximity correction, 46
Outdoor air, 236
Outsource, 94
Ownership, 229
Oxford English Dictionary, 33
Oxide nanoparticles, 11

P

Paclitaxel, 204
Parthenophobia, 189
Pathological liars, 186
Per se, 9, 12, 28
Percivall Pott, 117
Pernicious process, 78
Pharmacologia, 154
Phase change random access memory, 54
Phase shifting, 46
Phengophobia, 189
Phobophobia, 189
Photolithography, 60
Photonic crystals, 51
Photoresist, 46
Plague, 113
Policing Knowledge, 245
Politics of nanotechnology, 70
Polygamy, 186
Praxagoras, 109

Predinisone, 198
Prey, 42
pRNA, 208
Procarbazine, 198
Progression, 197
Promoter, 69, 197
Promotion, 197
Prophylaxis, 151
Protective, 251
Proto-assembler, 70, 76
Pseudospherical, 257
Pulmonary silicosis, 103

Q

Qu-bit, 52
Quantum, 258
Quantum bit, 52
Quantum dots, 89
Quantum numbers, 52
Quantum size effects, 29
Quantum well, 50

R

Radiation, 198
Raise the stakes, 251
Raphid, 190
Rationalized Plan, 83
Reality, 15, 89
Reality of nanotechnology, 14, 89
Renal replacement therapy, 217
Research monocropping, 82
Reticuloendothelial system, 202
Richard Wiseman, 115
Rights, 239
Roadmaps, 93
Robotics, 172
Rome, 110
Runner's high, 186

S

Salmonella typhimurium, 169
Satyriasis, 186
Scanning electron microscope, 50
Scanning microscopes, 49
Scanning near-field microscopy, 50
Scanning probe microscope, 4, 39
Scanning tunneling microscope, 49
Scanning tunneling microscopy, 59
Schistosomiasis, 103
Schizophrenia, 186
Scientific American, 34, 39, 41, 213
Scotophobia, 189
Self-assembled monolayers, 48
Self-consciousness, 248
Semiconductor nanocrystals, 257
Shaping the World Atom by Atom, 43
Shells, 28
Siderophobia, 189
Simplified, 252
Single electron transistor, 51
Single-walled carbon nanotube, 260, 263
siRNA, 209
Smokers, 186
Soil resisting, 55
Solid oxide fuel cells, 262
Solution- Liquid-Solid, 258
Soundbytes, 211
Space field, 71
Specific Design Guidelines, 68
Spendoholics, 186
Spin, 52
Spin logic, 3
Sponsor, 69
Sponsoring state, 71
Staphylococcus epidermidis, 146
Star Wars, 73
Stephen Bradwell, 115
Sticky fingers, 8
Sticky fingers problem, 4
Stone Age, 100

Streptococcus, 141
Sulphonamides, 134
Sun Microsystems, 64
Super thermites, 91
Surface acoustic waves, 57
Sweet tooth, 186
Synthetic, 248

T

Techno-economic paradigms, 86
Technological Revolutions and Financial Capital, 86
The Apologie and Treatise, 115
The Diamond Age, 89
The Enigma of Japanese Power, 84
The Future and Its Enemies, 184
The Society of Mind, 158
The Temperature in Diseases, 120
Thomas Young, 118
Thorazine, 134
Thrillseeking, 186
Tiger, 81
Top-down, 4, 45, 46
Top-down methods, 7
Top-down route, 226
Top-down techniques, 60
Tout court, 42
Tragedy of the commons, 234
Triskaidekaphobia, 189
Try to stop us, 252
Trypanosoma cruzi, 181
Tumour necrosis factor, 199, 202
Tumourigenesis, 196

U

Ultraviolet nano-imprinting technique, 47
Unbounding the Future, 160
Universal assembler, 5
Uploading, 63
Upside benefits, 251
Urticaria, 187
Utopian ideals, 61

V

Valium, 134
Venesection, 115
Vinblastine, 198
Vincristine, 198
Viruses, 8
Vis medicatrix naturae, 107
Vision, 15, 89
Vision of nanotechnology, 14
Vitamins, 68

W

Wagerphilia, 186
Weapons proliferation, 72
Wet, 28
Wet route, 226
William Cheselden, 116
Workaholics, 186
World War II, 83

X

Xenophobia, 189
XXM Newton, 47

Z

Zoophobia, 189